Botanical Illustration:

Preparation for Publication

By Noel H. Holmgren
& Bobbi Angell

The New York Botanical Garden
Bronx, New York 10458, USA
1986

Botanical Illustration

Published by The New York Botanical Garden, Bronx, New York 10458, U.S.A.

Issued 29 April 1986

Library of Congress Cataloging in Publication Data

Holmgren, Noel H.
Botanical illustration.

Bibliography: p.
Includes index.
1. Botanical illustration. 2. Botanical literature—Publishing. I. Angell, Bobbi, 1955– . II. Title.
QK98.24.H65 1986 580′.22 86-2548
ISBN 0-89327-272-8

Printed by Allen Press, Inc., Lawrence, Kansas 66044, U.S.A.

Cover illustration: *Mimosa irrigua* Barneby

Foreword

The quality of the illustrations that appear in any scholarly publication is a major concern to the editor and the author. Not only must the editor know how to evaluate illustrations, how they are reproduced, and what can be achieved with the materials and processes available, the author must also be informed about the production and evaluation of artwork suitable for publication. A synergistic interaction between artist, author, editor, and printer is required in order to achieve results of the highest quality.

Botanical Illustration is the result of the cooperation between Noel H. Holmgren and Bobbi Angell in the production of BRITTONIA. While evaluating, in conjunction with the associate editors, the illustrations submitted by authors for publication in this journal, Noel and Bobbi have encountered a variety of problems. Finding no single publication which addressed specifically the subject of botanical illustration, they concluded that a systematic account of their own might fill a gap. Three years ago Noel was asked to help re-write the "Plant Sciences" section in the *CBE Style Manual.* This catalyzed the writing of *Botanical Illustration.*

Botanical Illustration's purpose is "to provide information on how to produce acceptable artwork for publication" and is directed at "the student or non-professional artist who wants to upgrade the quality of his or her work, to the botanist whose report is being illustrated, and to the editor who may find some useful directives here." We hope that this contribution will enhance communication between artists, authors, editors, and printers employed in the production of high quality scholarly publications. Comments and suggestions for amplification or modification will be welcome.

María L. Lebrón-Luteyn
Curator of Publications
The New York Botanical Garden

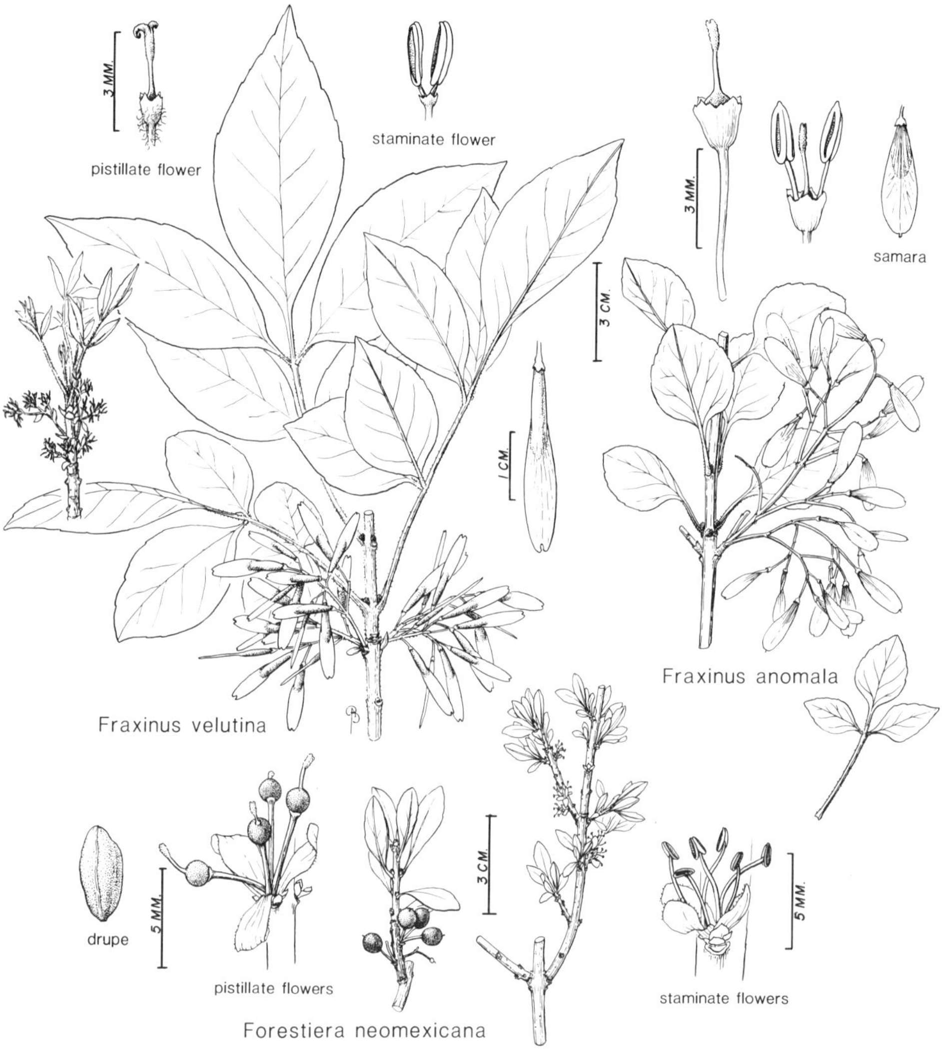

Illustration of three species of Oleaceae from the Intermountain Region (Page 343). *In*: A. Cronquist, A. H. Holmgren, N. H. Holmgren, J. L. Reveal & P. K. Holmgren. 1984. Intermountain flora: vascular plants of the Intermountain West, U.S.A. New York Botanical Garden, Bronx.

Preface

This book is based on our accumulated experiences in botanical illustration preparation, taxonomic research, and journal editing. Through answering questions from aspiring artists, dealing with poorly prepared illustrations from botanists who should have hired an artist and with beautiful drawings from those who did, and from experiencing our own mistakes and achievements, we have adopted certain techniques that work for us. We realize that there are many ways of producing good artwork and some techniques and tools may be better than the ones presented here. We hope to encourage a free flow of ideas and techniques that may be unknown to us.

Botanists, artists, editors, and publishers have in one way or another motivated us to write this book. We gratefully acknowledge those we have worked for and with; their influence and expectations helped to develop our skills. Our shortcomings as writers were improved by the invaluable help of the people who critically read the manuscript and offered suggestions. For this help we wish to thank Arly Allen, Christiane Anderson, William R. Anderson, Rupert C. Barneby, William C. Burger, H. David Hammond, Patricia K. Holmgren, Jeanne R. Janish, Sue Keller, and María L. Lebrón-Luteyn. We also thank Carol Gubbins Hahn for the use of her artwork and notes on her technique. Elsewhere, the artwork is by one of us. We thank the individuals and publishers who have granted permission to use copyrighted material; their use has been acknowledged individually in the legends. Our most special thanks go to Pat and Rob.

Noel H. Holmgren
Bobbi Angell

Table of Contents

Introduction

Well-ordered presentation is the sign of a well-ordered mind.
—Alfred A. Blaker (1977)

An illustration in a scientific publication should serve to enhance the text by clarifying concepts and by providing visual orientation; it should be an essential part of the publication. In botanical publications a plant illustration is especially useful in providing a clear, unambiguous image of a species. It can clarify key characters and descriptive phrases. When accurately done with wisely chosen detail it transcends in value the descriptive part of the text, especially for the foreign language reader. The importance of the saying, "a picture is worth a thousand words," cannot be overemphasized.

There is much creativity and conceptualizing involved in botanical illustration, whether it be a diagnostic drawing of a plant, a distribution map, or a simple graph. This is why we prefer to use the term "artist" rather than illustrator.

Illustrations in botanical publications range in quality from first-rate works of art to mediocre and even sloppy scribblings. In terms of usefulness, illustrations range from being the most important part of the article to a useless waste of costly publication space. Many poor illustrations result from a lack of knowledge of publication requirements, a lack of communication between artist and botanist or, perhaps just as often, from a lack of know-how rather than a lack of talent or skill.

Our main objective is to provide information on how to produce acceptable artwork for publication. To this end we discuss the limitations (and the potentials) of preparing art for the printer, the understanding necessary between the botanist and the artist, the availability and use of tools and materials, the kinds of illustrations needed, and how to care for the illustrations in shipment and storage. Our discourse is restricted to the preparation of botanical illustrations especially for systematic botany publications. The emphasis is on basic techniques and materials, so that a good quality illustration can be made with the least effort and the greatest cost-effectiveness. We stress the use of high quality and long-lasting materials because artwork often retains its usefulness long after it has been published. We are directing this to the student or non-professional artist who wants to upgrade the quality of his or her work, to the botanist whose report is being illustrated, and to the editor who may find some useful directives here.

Unlike the written text, an editor usually does not correct flaws in submitted illustrations and the printer cannot substantially improve the appearance. Poor quality art delays publication, costs more to reproduce, or may be a reason for rejection. Once published, bad illustrations not only reflect poorly on the study and the author, but on the editor, the publication, and of course the artist. Raising the quality of illustrations and reaching for high standards will improve the quality of the entire journal.

Kinds of Illustrations and Their Uses

Illustrations, as defined here, include drawings of plant species, graphs, maps, and photographs as used to enhance the written word. As seen by the printer, illustrations fall into two basic categories: line and halftone art.

Line art includes pen and ink drawings of plants, graphs, and maps, in which any tonal qualities are achieved with stipples and hatched lines. Line art is always drawn and printed in black and white with no shades of gray.

Halftone art (also called continuous-tone art) includes photographs, paintings, and pencil drawings. The originals have continuous gradations of tone that may range from solid black to paper white and the printed results yield a similar effect. A halftone requires more steps in preparation for printing than line art and is therefore more costly to reproduce.

Color reproduction is expensive and most journals require authors to bear the extra cost. In limited-circulation botanical publications color reproduction is rarely used.

The subject matter often dictates which kind of art should be used, but the cost of the artist's time and the cost and quality of the printed results are factors that must be considered as well. Efforts to make high resolution prints of finely detailed line drawings are meaningless if poor reproduction methods are used or if the work is printed on poor quality paper.

The type of paper used in the journal noticeably affects the end results. A low quality paper is the pulp paper used in paperback novels, newspapers, and some journals. On such paper, line drawings must be simple and uncluttered, with relatively bold lines and coarse stippling. Photographs must be suitable for a coarse screen in which the printed dots will be detectable to the unaided eye. Dots on low quality porous paper enlarge as the ink is absorbed. When the dots are closely spaced the ink spreads from one to the next, resulting in uneven and darker tones than intended. This is also true of lines, although they do not appear to enlarge as much.

Better quality papers yield better reproduction of all artwork. The finest papers are usually coated with white clay or an acrylic substance to provide a smooth surface. Quality paper can be matte or glossy coated depending on the final processing steps. Matte paper provides a dull, non-reflecting surface and is preferable for line art. A glossy surface is best for finely screened halftone art.

Working Relationships

Communication between Artist and Author

An important part of producing successful botanical illustrations has to do with communication between artist and author. Sometimes the author (botanist) and the artist are one and the same. More often an author and artist collaborate, and then there should be a clear understanding of what the author wants the artist to do and what the artist is capable of doing. It is helpful when the artist has some training in botany; it is also useful if the author has some understanding of what goes into making a good drawing. The artist should know enough botany to be able to understand the manuscript and to communicate with the author in technical terms.

If the author does not already know of an artist, the following suggestions may be helpful. A university art department may be a good place to inquire, or a want ad may be placed in the school or local newspaper. Although it is preferable to deal with an artist in person, it may be necessary to conduct business through the mail. The Guild of Natural Science Illustrators, Inc. (P.O. Box 652, Ben Franklin Station, Washington, DC 20044) can provide names of artists. Curators in larger herbaria can be consulted for names of artists who do free-lance work.

Until an artist and author get used to working with one another, care should be taken to make sure that the end result will be satisfactory for both. With experience, most guidelines become second nature and the competent artist can proceed with a minimal amount of discussion. When working together in the beginning, however, an artist and botanist should develop as much communication as possible.

The artist should know the intended use of the illustration, and whether or not it is to be published along with others, to which it should conform in style and content. The artist should also be aware of the journal's specifications, such as page and column dimensions and the editor's preferred reduction. If the author is not sure of its intended use, the artist should work at a standard size. Many journals have a printed area of five by eight inches (see section on size and proportion).

Sufficient time should be scheduled for sketching and reviewing of a preliminary version and for final preparation. If the author leaves the illustration until the last minute, he faces the risk of a sloppy, rushed job.

The author and artist should discuss time and cost limitations if they are to determine the amount of detail to be drawn. Often an artist can work within a botanist's budget, preparing a simpler or more detailed drawing depending on the funds or time available.

Planning the format in advance saves both the artist and author trouble later on. The author may have very specific ideas of how the details should be laid out but more often these decisions are left up to the artist. If the author has specific ideas about content, layout, sizes, and views, he or she may want to provide a rough sketch. This is helpful especially if the plant has a characteristic habit that may not be evident in a herbarium specimen, such as decumbent branches. The author should prepare a list of what is to be drawn, and when possible provide copies of sample drawings of related species that can help clarify what is expected of the artist. The diagnostic characters should be explained and the artist should be given a copy of the manuscript with key and description if possible. Measurements, proportions, and shapes that the botanist feels are representative of the species, as well as standard views and aspects for a genus or family should be explained to the artist.

The artist should be given adequate material to draw from, which may include herbarium specimens, pickled material, live material, and photographs. The artist may find unmounted specimens easier to work with, as they can be turned over to reveal the best side. The artist must know whether to illustrate a specific plant specimen (such as the holotype) or to represent a generalized ideal from more than one specimen. An experienced artist can pick the material easiest to work with, but the choice must coincide with what the author thinks is most representative and important.

As an added measure of accuracy the artist should record which details were taken from which specimen to enable the author to cite the sources. This can be written on the back of the illustration; a specimen can be referred to by the collector's name and field number. The author should likewise retain a record of the specimens and instructions given to an artist. The author should be assured that the artist knows how to handle specimens and has the proper storage facilities for them.

The author should have enough specimens ready and projects assigned in advance so the artist does not waste time having to return for more. To the free-lance artist, time is money.

Maps and graphs require some of the same considerations. Information needed to prepare a map includes geographical features, data points, legend material, and names of identifiable features. A sketch map with explanatory notes and a list of (correctly spelled) names is extremely helpful.

The artist should prepare a clear, non-ambiguous pencil drawing for the author to review. If the sketch is covered with tracing paper the author can mark any corrections or comments directly on the cover. Another possibility is to make a photocopy of the drawing so that the author can mark it up without restraint. Good photocopy machines can pick up #2 pencil lines. If working through the mail, the artist can mail a photocopy without much worry about abuse or loss in transit. The author should examine the sketch thoroughly, for it is ultimately the author's responsibility to catch errors (even though the artist may be blamed).

The author should recognize the importance of the artist's contribution and give proper credit in the publication's acknowledgments, even if the artist has signed the plate. Artwork is made under the author's direction, but often the author makes only a few suggestions and provides poorly prepared and inadequate herbarium specimens, leaving to the artist the problems of execution. Much research, training, and skill goes into good artwork. Artists achieve prominence and reward from having their name in print, just as authors do. Often readers will consult the illustrations in identifying plants or comparing species before resorting to the author's written keys and descriptions.

U.S. Copyright Law

An understanding of what the U.S. Copyright Law covers should be of interest to artist, author, and editor (or publisher). The law was substantially changed in 1978, generally in favor of the artist.

Exclusive Rights.—Under the present U.S. Copyright Law, in effect since the beginning of 1978, exclusive rights and protection of such rights by law begin with completion of the artwork without the necessity of registering or publishing it. These exclusive rights include the right to (1) reproduce the artwork, (2) prepare derivative works (modification or adaptation) based upon it, (3) sell and distribute (publish) it, and (4) display it publicly, including photographic slide shows and lectures. A fifth right of copyright ownership is the exclusive right to perform a work publicly (music, plays, poems, etc.).

Work for Hire.—An understanding of the concept of "work for hire" in the present Copyright Law is especially important to the free-lance artist. With a work made for hire, the botanist, employer, or whoever commissioned the work is considered to be the initial owner of copyright as if the employer created the work himself. To be considered a work made for hire it has to be created by an employee within the scope of his or her employment, or on special order or commission if (and only if) both parties sign a written agreement that it is a work for hire. If no such agreement has been made, it is not considered work for hire and the free-lance artist owns the rights.

Control of Copyright.—Ownership of the copyright is completely separate from ownership of the actual artwork. Unless there is an agreement to the contrary, the transfer of copyright does not convey ownership of the work of art, just as the transfer of ownership of the work of art does not convey the rights of copyright. When you buy a painting you do not automatically acquire reproduction rights.

If an artist has any interest in maintaining rights to the artwork, he or she should avoid the use of work-for-hire contracts. It is best that the author and artist understand the intended uses of the artwork and agree on terms in writing as soon as, or before, the artwork is completed. In scientific publications the copyright law is generally less important to the artist than in more commercial fields, and with considerate attitudes on the part of both the author and the artist, problems with copyright infringement can be avoided. It is common for a botanical illustration to be used more than once, perhaps first in the publication of a new species, then in a monograph or flora or both. The uses may occur over a long period of time, and negotiation with, or locating, the artist may be difficult at a future time. For most scientific illustrations, the artist would have little use for the artwork beyond portfolio and display purposes, and should consider leaving the illustrations with the institution with the understanding that he or she will be contacted if further use is made of the artwork. In due respect to the botanist, the artist should not use the artwork in any manner competitive with the author's use or intended use and should agree to request permission and credit the published source. The agreement should be in writing to avoid any later problems.

In situations where the artwork may be used for profit, or used in ways that an artist may wish to control, the artist should include copyright notice on the artwork, register the copyright in his or her own name with the U.S. Copyright Office, and design a contract that transfers only specific rights, such as "First

North American rights." Copyright notice includes the word "copyright" or the symbol for it (©), the artist's name, and the date (year is sufficient). This can be placed anywhere on the illustration, including the margin or back. With all the rights of copyright maintained by the artist, the author cannot make subsequent use of artwork without negotiation. The price of the artwork may vary in relation to the rights transferred—the more rights surrendered the higher the fee the artist should charge.

Transfer of Copyright Ownership.—Each of the rights of copyright can be transferred and owned separately. Under the old law (1909 Copyright Act), the publisher of a collective work (journal or book) automatically acquired complete copyright ownership of individually submitted contributions and this still applies in considering the ownership status of pre-1978 publications. With the current law, however, the publisher must have a written transfer agreement for each contribution, signed by the author, to hold the exclusive right to it. Without such an expressed agreement the publisher is restricted to copyright for the collective work as a unit with only the rights to reproduce and distribute the artwork as part of that collective work.

The concept of a collective work can be projected to another level. A scientific paper consisting of the author's text and any number of illustrations contributed by a free-lance artist is considered a collective work. The law states that the copyright in each contribution is separate from the copyright in the entire collective work. If the artist and author have not made a written agreement, either stating that the art is a work for hire or transferring the copyright, then the artist retains all the rights even when the author transfers his or her rights to the publisher.

Permissions.—With transfer of the copyright, the publisher has been given the right to reproduce and distribute the work (which of course is the author's intention in the first place) and only if the signed agreement so states, the publisher has the right to grant permission to others to reuse parts of it on a nonexclusive basis. It should be stated in the agreement that the publisher grants this right with the stipulation that the source, artist, and author are cited. This use is nonexclusive in that with this permission the requester is not in a position to complain if others are also granted the same rights from the copyright owner. In other words, the requester does not have statutory right to sue for copyright infringement. The requester should also be expected to seek permission from the original copyright holder (artist or author), not only as a matter of courtesy, but as an avoidance of the risk of legal infringement, due to ambiguities in the written transfer agreement.

For More Information.—Some good sources for further details on copyright, such as term, termination, registration, notice, fair use, securing permissions, and infringements suits, are Crawford's "The Visual Artist's Guide to the New Copyright Law" (1978) or Johnston's "Copyright Handbook" (1982). Artists can obtain a copyright information kit from: The Copyright Office, Library of Congress, Washington, DC 20559. The Graphic Artists Guild annually publishes *Pricing and Ethical Guidelines,* which contains standard contracts to copy or adapt.

Tools and Materials and Their Uses

The different kinds of tools and materials available for the preparation of illustrations are often expensive and the use of some of them may seem obscure. There are some tools that will save time and improve the quality of work and are well worth their cost, whereas others serve only to save a little time and may end up being stored away as useless gimmicks.

Basic tools include a light box, pencil (or lead holder), sharpener, technical pens, ruler, miscellaneous templates (at least for assorted circle, square, and triangle sizes), T-square, right-angled triangle, compass, dust brush, and hand-held cutting tool.

Hardly anyone reads a catalog from cover to cover, thus many useful tools remain undiscovered or misunderstood. The following is a list of items of which we are aware and a discussion of their use.

Furniture

The standard furniture for art preparation includes a drawing table (or simply a drawing board on a desk or tabletop), a chair, storage facilities for tools and supplies, and a good adjustable lamp.

A drawing surface that can be tilted at an angle provides a vantage point for viewing artwork without distortion and allows easy access to top and bottom without back strain. A swivel chair with an adjustable height and back support and easily accessible supplies and tools take the drudgery out of art production. Some artists find that it saves energy to have the surface at bar height for easy exit and entrance. During long sessions, periodic readjustment of one's chair height may prevent back or neck ache.

The advantage of a good light source is important. Poor lighting can result in eyestrain and headache. Fluorescent lights strain some people's eyes with their imperceptible flicker, and the persistent hum of the transformer may take its toll on other nerve centers. An incandescent light with a 60 or 75 watt bulb will alleviate these problems, although it does radiate heat. Combination fluorescent/incandescent lamps are available. An adjustable-arm lamp is most useful, anchored on a corner of the desk. For right-handed artists the light source should be from the left side to avoid casting shadows from the drawing hand, and vice-versa for left-handed artists.

Microscopes

A dissecting microscope with an illuminating fixture is a necessary piece of equipment for artists working with details of plant specimens. It should have

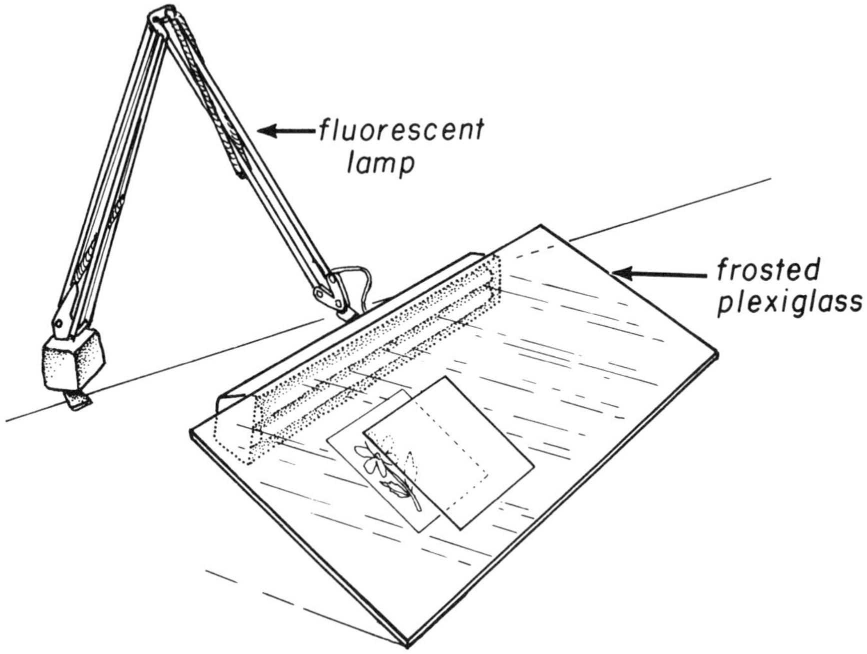

Fig. 1. Makeshift tracing board using a fluorescent desk lamp and plate of glass or sheet of plastic.

a long horizontal arm on a counter-weighted base to allow free movement of herbarium specimens beneath the lens. It should range in power from 10× to at least 30×, providing an overall view and resolution of fine detail. If an ocular micrometer is used it should be calibrated by the artist with a ruler, especially if the microscope has a zoom lens. The microscope should be positioned so that the artist can easily switch the eyes from the lens to the drawing without neck strain.

The microscope light source should come from the side to provide shadows for resolution of detail and depiction of depth. A choice of black and white viewing surfaces, consisting of boards, plastic sheets, or glass plates, is useful for highlighting translucent flowers or white hairs.

Compound microscopes are necessary for artists working on cellular details of mosses, liverworts, etc. They are most useful if outfitted with a drawing tube, which allows the artist to transfer outlines onto a sheet of paper (see section on drawing tube).

Light Box

A box with a frosted white glass top and inside light source, used for transferring sketches onto bristol board for final rendering, is an indispensible item for many artists. Of the commercially available light boxes, those with two parallel fluorescent tubes provide reasonably even illumination. The artist may want to choose one that can be tilted at an angle for greatest comfort. In addition to desktop and portable light boxes, inexpensive temporary options are available. The makeshift method of holding work against the window pane with daylight as the light source works, but has its obvious disadvantages. A better and not too expensive substitute is a sheet of $\frac{3}{16}$ or $\frac{1}{4}$ inch thick, 15 by 20 inch (38 × 50 cm), translucent white plexiglass propped on stacks of books with a light aimed underneath. A fluorescent desk lamp (with straight tubes) on an adjustable arm that can be bent down to the desk surface with the reflector tilted so the tubes shine upward, with the plastic sheet against it, works well also (Fig. 1).

Papers, Boards, and Drafting Films

Bristol Board.—In artist's vernacular, fairly thick and heavy paper is called "board." Bristol board, a generic term, is the most frequently used drawing paper for botanical illustrations. It ranges in thickness from one- to five-ply. Two-ply is the best to use for its firmness and strength and for its light-box transparency. It can even be traced on over another two-ply sheet on a light box.

Bristol board is available in two surface types: vellum finish (kid, matte, or medium surface), which has a barely perceptible rough surface, and plate finish (glossy or high surface), which is smooth. Some brands have a supposed right side, which is the side from which the embossed seal can be read, but most boards are the same on both surfaces. The vellum finish is good for pencil, charcoal, or crayon work, but these media are rarely used for botanical illustration. The plate finish is best for pen and ink work. Bristol board comes in pads of various sizes or in large sheets, usually 23 by 29 inches (58 × 74 cm), which can be cut down to appropriate sizes, and are less expensive than the pads.

A poor quality paper does not take ink very well; the ink will sometimes run thin and consequently gray, or its flow may feather out. A high quality paper takes pencil and ink well and withstands repeated erasures without damage to the surface, and self-adhesive tapes, films, and letters can easily be repositioned without breaking the surface.

Acid-free 100 percent rag paper is recommended for any work that is expected to last. The long cotton fibers are less easily disturbed by the pen point or eraser. Strathmore is a commonly available brand of rag bristol. At the top of their line is the "500" series, which is 100 percent rag. The "400" series is part rag; it is a good surface upon which to work but will not hold up over the years. There are other good brands, but the artist should inquire into their rag content and check the quality of a pen line before using it for publication.

Tracing Paper.—Tracing paper or tracing vellum is available in several weights and transparencies; the most finely grained paper is the most transparent. Coarser papers are adequate for most sketching and layout work. There are many brands available. Tracing paper is primarily used as an intermediary in transferring outlines and is not satisfactory for finished work.

Vellum.—Good quality rag tracing paper is usually called vellum. It is much more expensive than standard tracing paper, but is considerably more durable. It is recommended as a protective covering sheet, as it is semitransparent and relatively sturdy to survive the editor's and printer's handling, and it will not stain the underlying sheet in storage.

Drafting Film.—A frosted semitransparent drafting film is suitable for final artwork, and can be used instead of bristol board. Drafting film (also called mylar) is 100 percent polyester resin and will not warp or stretch under normal use. It is waterproof, resists tearing, and is long-lasting. Final artwork can be done directly over a preliminary sketch, eliminating the need for a light box. It is frosted on one or both sides and such a surface takes ink well. Only one side should be used, otherwise the printer will have problems photographing through the thickness. It also takes pencil and withstands repeated erasures. Ink work is easily correctable with a plastic eraser, scalpel, or single-edge razor blade. A drawback is that the film's abrasive surface wears down pen points, making it difficult to keep a consistent line. Also, ink will not adhere to grease marks, so the surface must be kept clean. When working on it a clean piece of paper should be placed between the drawing hand and the film. Rubber cement thinner and a clean cloth can be used to wipe away any grime or skin oils before beginning work.

Common brands of film are K&E Herculene and Cronaflex, which are available in packages of single sheets or large rolls.

Drafting film should not be confused with acetate or cellophane. These lesser quality films are acidic and should not be used for artwork or as cover sheets. Drafting film also should not be used as a protective cover if adhesive-backed applications, such as press-on letters, have been used as it may lift them off before the illustration gets to the printer.

Graph Paper.—Gridded tracing paper, also called cross-section or quadrille paper, with four to ten squares to the inch or centimeter is useful for scaling drawings up or down. Its intended use is for graphs, but it is useful also for balancing and aligning drawings. Blue or light green ruled lines on graph paper will not reproduce and can be used for final artwork. Gridded bristol board is available.

Poster Board.—14-ply poster board is a good inexpensive mounting board. It is often called crescent board after the name of one of the popular brands. The board should be used for mounting artwork that might be handled roughly, and for all artwork with pressure-sensitive labels that may peel or chip with rough handling.

Pencils and Pens (Fig. 2)

Pencils.—Pencils come in different degrees of hardness and diameters. The medium range in hardness (H, 2H) is preferable, as its line is clean, easy to read, and easy to erase. Harder leads (3H or more) will engrave or, in humid weather, even gouge the paper. They are more difficult to erase, and will not photocopy well. Softer leads smudge too easily.

Mechanical pencils with refillable leads are available for leads in all degrees of hardness and in thickness ranging from 0.3 to 0.9 mm. The lead is advanced by a twist of the top or by a push of a button, eliminating the need for sharpening. They are useful for field sketching and quick notes.

Lead holders are a type of mechanical pencil designed to fit into a lead pointer for needlepoint sharpening. They can be sharpened faster and finer than standard pencils. Because only the lead is replaced as it wears down, the lead holder lasts indefinitely at the same weight and feel within the artist's hand, unlike normal pencils that range from long new ones to consumed stubs.

A lead pointer should either be fixed to a desk or be heavy enough that it

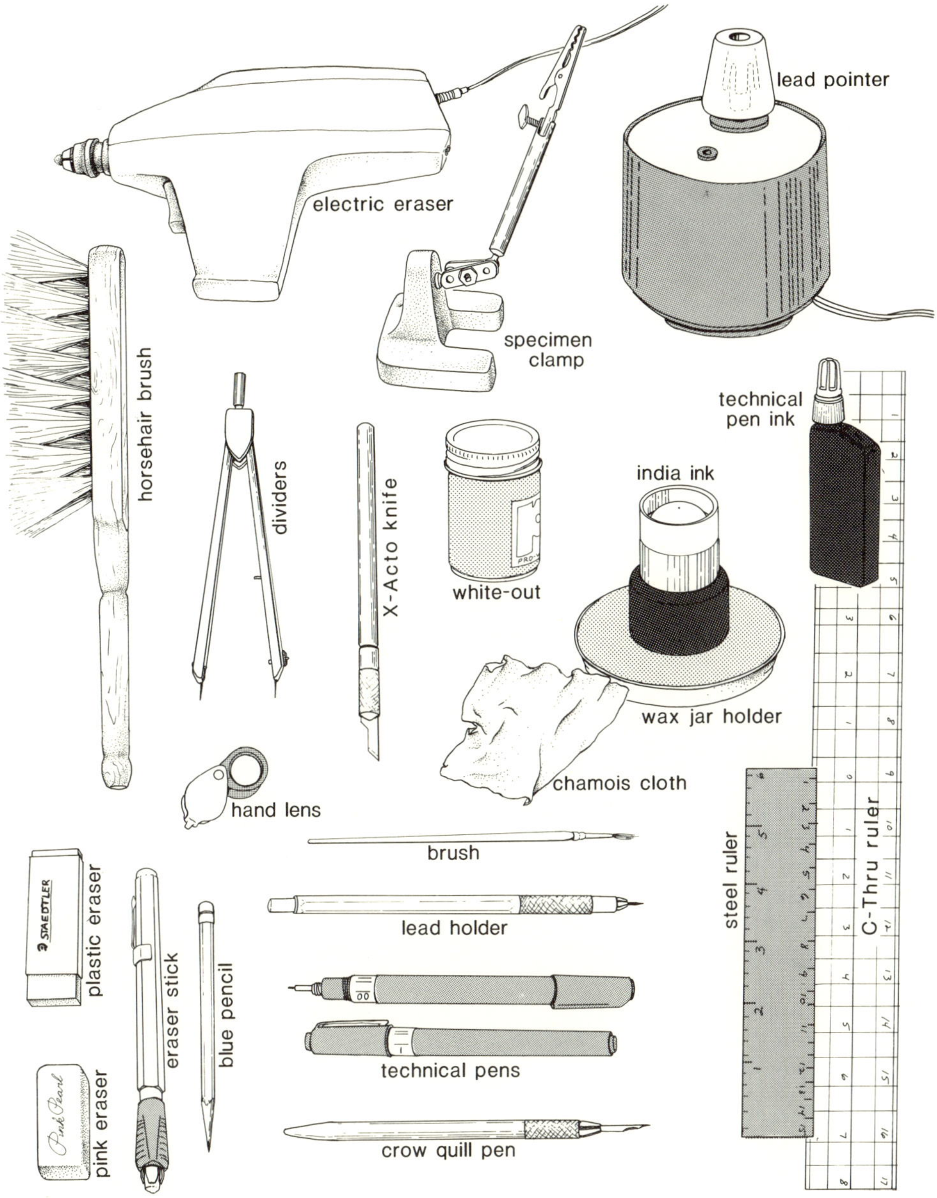

Fig. 2. Basic tools for pen and ink illustrations.

does not move. An electric lead pointer sharpens the point fast without disrupting work or shaking the desk as the manual ones do. It may seem excessive, but once you have one you will never bother with a dull point again.

Guidelines or notations done in light blue (non-photographic) pencil or pen will not register in the high-contrast photographic film used in line-art reproduction. But they are often difficult to erase and should be restricted to the margins on work destined for exhibition. "Blue-zip" is a brand of blue-colored magic marker that can be removed with its companion "eradicator stick," which fades the blue line almost to invisibility.

Technical Pens.—Since its introduction 30 years ago, the technical pen and the specially formulated inks used with it have improved considerably. The more commonly available brands are Mars, Faber-Castell, Koh-I-Noor Rapidograph, and Keuffel & Esser Leroy. Technical pens come in widths from 6×0 (000000) to 6 point, for line work, lettering, and scriber systems. Line width numbers are somewhat arbitrary and the scale differs slightly among brands. They are not the same as the printer's point sizes (see section on figure labels), but among the brands listed above, 0 equals about 1 point ($\frac{1}{72}$ inch), 2.5 equals about 2 points, 3.5 about 3 points, and 6 about 4 points ($\frac{1}{72}$–$\frac{1}{18}$ inch or 0.35–1.4 mm).

Only ink that is specifically formulated and labeled for technical pens should be used with them (see section on inks below).

Technical pens used frequently are generally reliable, but those used infrequently will inevitably clog and need cleaning. Humidifiers are available for storing pens. Faber-Castell and Rapidograph have caps specially designed to keep the ink from drying and clogging the flow tube. If a fine-point nib is even minutely clogged, it can filter out the large pigment particles from the ink and produce a gray line. To start the ink flow of a reluctant pen, shaking it with a gentle horizontal motion and then drawing it across a scratch pad will usually start a smooth flow. With more stubborn pens, vertical shaking with cap in place and gentle tapping on the desk top may free up the flow. Harsh tapping could damage the needle inside the pen. Wiping the tip with a damp rag or dabbing it in water may help if shaking is not enough. If a clicking movement cannot be heard, the pen must be cleaned before use. If you anticipate not using a pen for a while you should clean it when a simple water flush will still do the job. Allow it to dry thoroughly before using. Specific cleaning fluids are good for thoroughly clogged pens, but a few drops of liquid dishwashing soap with ammonia added to warm water works also.

An ultrasonic cleaner, a small, liquid-filled tub, will clean a badly clogged pen in half a minute. Jeweller's models, if you can find them, are often less expensive than the artist's models. Some art supply stores will ultrasonically clean your pens for you.

The standard stainless steel points last a long time with regular use, but if you use abrasive frosted drafting film you may want to buy the longer-lasting jewel or tungsten carbide points. While they may cost three times as much, they may last three times as long if properly taken care of.

Ruling Pen.—The ruling pen (Fig. 3) is not commonly used anymore. It is a forerunner of the technical pen, with a caliper-like adjustable gap for line width. A drop of ink is placed within the gap and is drawn out between the points. Too much ink may cause blobs, too little ink will not complete the line.

The railroad pen consists of paired ruling pens with an adjustment controlling the distance between the two. It is used for drawing parallel lines, as for roads on a map.

Crowquill Pen.—Many artists prefer to use drawing (dip) pens over technical pens. They are available in a great range of styles, sizes, and flexibilities. Hunt 102 is the standard "crowquill," but other nibs, such as Hunt 104, 108, and Gillott 850 and 659 are commonly used. You should experiment with different kinds and you may end up with more than one favorite pen for different effects. The flexibility of the point and the size of the hold are factors in performance.

Holders are available ranging from inexpensive lightweight plastic holders to metal ones with adjustable jaws that hold the nib.

Crowquills take a bit of practice, but are more flexible than technical pens in their range of usefulness as varying pressure provides different weight lines. New nibs must be broken in through gentle use until the point is slightly worn down. It is best to have two pens, a good sharp one for fine lines and a rounded one for stipples and bold lines. A small piece of chamois or cotton cloth and a small container of water are useful for cleaning the nib whenever you pause long enough for the ink to dry.

Brushes

A brush is occasionally useful for painting in black areas or correcting errors with white paint. India ink and white-out paint are hard on brushes, especially if the artist uses them intermittently without a thorough rinsing. It is best to have two inexpensive brushes, one for white-out and one for india ink in a fairly fine size such as number 0 or 1.

Ink

Intense black waterproof drawing ink is essential whether used in a technical pen or quill pen. Fountain pen ink is not suitable as it is not black enough and fades with time. If ink is not dense black the printer's photographic film may drop out some of the weaker lines, or extra effort may be needed to achieve good quality. For technical pens some of the brands of ink that work well are Rapidograph, Pelikan, Staedtler, and Higgins. For dip pens Pelikan black (tusche) drawing ink is recommended, but Higgins is good also.

You can select fast- and slow-drying inks for technical pens depending on whether you are working under humid or dry atmospheric conditions. The slower-drying inks are also better for small point sizes and conversely, the faster-drying ones are preferable for the larger point sizes. Rapidograph lists the inks they recommend for their pens in order from fast-drying to slow-drying: Rapid-O-Ink, Pelikan Type-T, Koh-I-Noor Rapidraw, Rapidograph Universal, Pelikan Special, regular Pelikan, and Rapidograph Plotter.

Old ink tends to separate, giving a dilute gray, and can clog technical pens as the large grains stay behind in the pen and a dilute ink runs out. If in doubt about the age of a bottle, buy a new one.

Nothing ruins a drawing (or desk, or floor) like spilled ink. Ink bottle holders are available to make bottles tip-proof, but they are hard to find. A homemade wax/jar holder (Fig. 2) is easy to put together. Melt paraffin or candle wax, pour it in a wide-mouth jar lid (peanut-butter jar lid works well), set a glass eye-dropper-stoppered ink bottle in it, and let cool. Sand paper can be glued onto the bottom of the mount to keep it from sliding. The ink bottle can be replenished as the ink is used. Pelikan no longer sells its ink in glass bottles, but Higgins still does. Empty glass bottles are sold separately in some stores.

Erasers

Erasers come in blocks, peel-off pencils, and long cylindrical refills for eraser holders and erasing machines (Fig. 2).

The white plastic eraser is nonabrasive and non-smudging. It is made of vinyl and was designed for drafting films, but it works well on bristol for removing pencil. It will not remove or damage ink, except when used wet on drafting film. It is available in several shapes. The pencil-shaped plastic holder with refills made by Mars Staedtler is especially useful. The eraser, as it becomes a round stub with use, can be carved into a round point.

Koh-I-Noor makes a yellow vinyl eraser, infused with erasing fluid, for removing ink from drafting film. It is available in a block or in electric eraser refills.

The pink eraser is good for removing pencil lines but may be slightly abrasive. When it becomes dirty and causes smudges its surface has become impregnated with skin oils or has hardened due to oxidation. It can be revitalized by removing the old surface with sandpaper.

The gum eraser, a large crumbly block, is used on bristol board to clean up pencil marks and smudges around finished lines. It is not harsh on paper and does not remove ink. However it is rather sloppy for general use.

The kneaded eraser is not recommended for pen and ink work as it can partially remove ink and can leave a greasy film on the paper that will not take ink. It is a soft, moldable rubber, best used for toning pencil and charcoal work.

An electric erasing machine can be fitted with different textured erasers for cleaning up pencil marks, smudges, inks, and typescript. The rapidly rotating cylindrical eraser removes marks that could not be removed by hand. The initial investment is soon forgotten when drawings are saved by cleaning up smudges and correcting mistakes. Thus the botanist can suggest corrections or changes in a final inked drawing with little fear of reprisal from the artist.

The gritty, gray eraser, made for the electric erasing machine, removes ink lines and blots by removing the surface of the paper, but the surface is left too rough to be reworked. The softer pink eraser is preferred, as it removes ink without perceptibly damaging the surface. It takes longer to remove ink than the gray eraser, but the paper can be re-inked as normal. Removing ink with the pink eraser may take a little patience but it can effectively salvage a drawing. An erasing shield, a thin, flat metal card with variously shaped slots, is helpful for accurate spot erasures.

The single-edge razor blade is another useful tool for removing ink lines and blots. Hold a shiny new blade upright and scrape with its whole edge, not just one corner, with light pressure over the blot from various directions. A sharp blade powders off the paper, leaving the surface as smooth as the original.

Another more accurate technique, using a scalpel or razor blade, is to cut a very fine line around the mistake and peel off the top layer of bristol. This works well when a line has gone too far, as in longitude lines extending beyond a map border. This procedure is for final cleaning up; the excavated areas do not re-ink well.

White opaquing fluid (white-out) is often used for making corrections. It can be brushed on as an opaque surface. Pelikan Graphic White gives a waterproof, non-glare surface. Pro-white and Q-white are also good. When applied too thickly, any of these fluids may crack or chip off. Apply them in one or more thin layers as needed. White paint is easily dirtied. When working near the application, place a piece of paper under the hand to keep the surface clean and white.

For large corrections a paste-over job may be preferable, using a clean piece of bristol glued on top of the original. With photographic reproduction of line drawings the shadows of the edges will not show. Self-adhesive typewriter correction tape is also useful for masking errors.

Measuring and Ruling Tools (Fig. 3)

Measuring Rulers.—The artist needs a good accurate ruler, preferably one marked in both inches (to at least $^1/_{16}$) and a centimeter scale (to the millimeter). A six-inch (15 cm) ruler is a convenient size to work with most of the time, but you should have a 12- and an 18-inch ruler as well. At least one ruler should have a beveled or raised edge for inking, to prevent wet ink from coming into contact with the edge of the ruler and running underneath causing that dreaded blot. The most useful ruler is a metal one with a strip of slipproof cork backing.

C-thru makes a transparent ruler with inches to the $^1/_{10}$ and metric to the millimeter. The 18- by 2-inch version has tenths of an inch marked in an overall grid which is useful for making parallel and perpendicular alignments.

A pica ruler may come in handy for some jobs involving printer's specifications. If an editor designs a specific space for an illustration, he or she will usually give the size in picas (see section on figure labels). C-thru makes a (nontransparent) Graphic Arts Ruler which also doubles as a cribsheet for publication information. It gives pica and inch calibrations and other information, such as proofreader's symbols, printer's rules (hairline to 12-point lines), percentage screens (10% to solid black), halftone screens (55 to 150 line).

Schaedler Precision Rules are used as standards in the graphic arts industry. The rules are hairline-thick for accurate measuring. Their two-piece set includes a 12-inch ruler with inches subdivided to $^1/_{64}$, and centimeters to 0.5 mm, and a pica ruler with point lines and agates.

A steel straight edge for cutting is necessary. Never use a plastic ruler for cutting, and it is best to avoid your favorite inking ruler, be it metal or plastic. One nick will render it useless.

A triangular scale, a ruler that is concave-triangular in cross section, places the ruled markings flush on your copy for ticking off accurate intervals.

T-Square.—A T-square is a long, straight-edge tool with an offset end mounted at a right angle to the edge (Fig. 3). When run along a drawing board, it can be used for drawing horizontal lines, aligning drawings and lettering, and as a guide for triangles and lettering stencils. Its length should be at least six inches shorter than the width of your drawing board for movement up and down without knocking off pencils, pens, ink, etc. A T-square with a clear plastic edge allows you to see what is underneath, but the edge turns yellow with age and nicks easily. An all-metal one is recommended, as it will last forever.

Parallel/straight edge drawing boards have an attached moveable bar that is kept parallel by a system of cables and pulleys. The boards are available in different sizes. The parallel bar takes up the full width of the board and always takes up space at the top of your drawing surface, so it is not recommended unless you need to do a lot of parallel line work.

Triangle.—Triangles (Fig. 3) are used with a straight-edge for aligning or drawing lines at a perpendicular or other angle. They are made in different sizes and angles. The most useful one has a beveled edge for inking.

Templates.—For penning consistent circles, squares, triangles, ellipses, etc., there is a myriad of templates available in art and stationery stores. For inking, the template should be elevated from the surface of the paper. You can elevate the drawing edge by attaching narrow strips of tape onto the underside or you

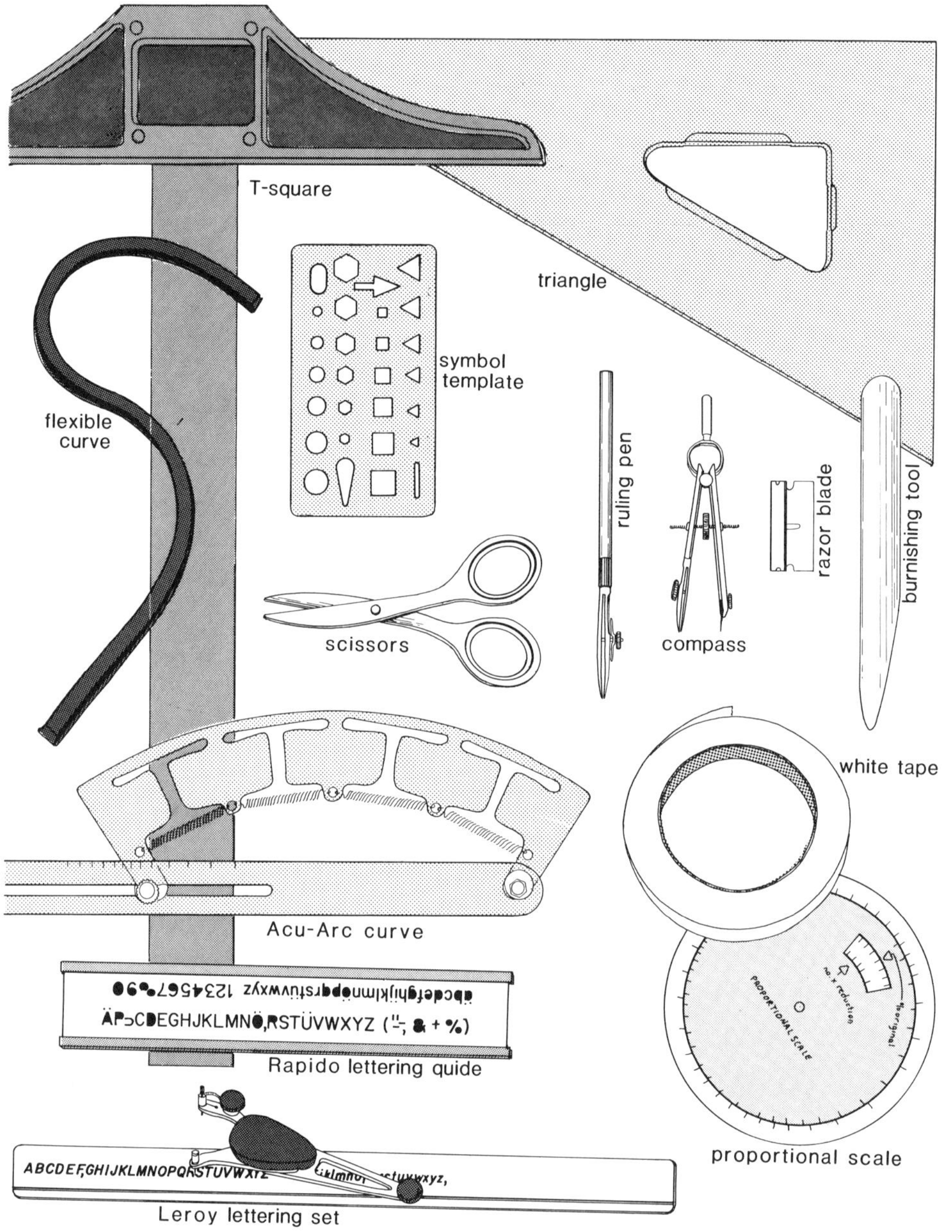

Fig. 3. Basic tools for graphics.

can use a "template ink riser," an unpretentious little machine that simply makes dimples in the plastic.

Curved Rules.—A collection of French curve rules (flat plastic paisley-shaped tools) with raised edges for inking may come in handy.

The Acu-Arc ruler (Fig. 3) has a 12-inch ruling edge that can be easily adjusted to curves of any radius larger than 6¾ inch (17 cm). It is especially useful in map work when you must ink curved longitude lines and boundaries.

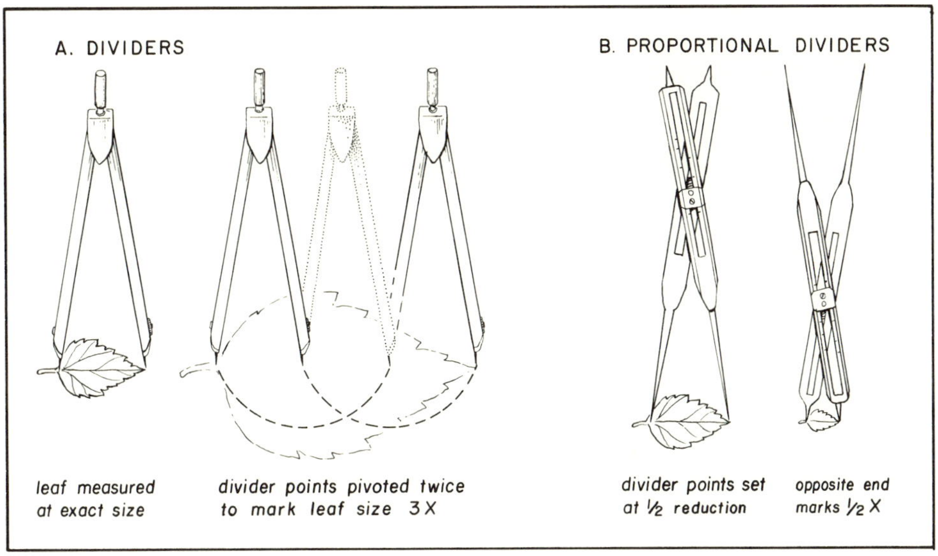

Fig. 4. The use of dividers (left) and proportional dividers (right) in scaling up or down.

A flexible curve (Fig. 3) can be bent to any shape for inking curved lines. It consists of a lead core encased in plastic.

Compass.—For drawing circles and arcs, a compass (Fig. 3) is needed. Some have interchangeable attachments for holding pencil leads, technical pen nibs, and ruling pens. The center point should be sharp and fine so large holes will not be left in the drawing paper. Dime store models are too coarse.

Dividers.—Dividers (Fig. 2) are used for transferring measurements from the source to the drawing. The subject is measured with the points of the arms, then marks are made on the paper with small pricks. The dividers can be used in scaling up by rotating the points of one leg around the other two times, three times, or more (Fig. 4). Enlargements are as accurate as the first measurement, and inaccuracies are magnified. A professional grade pair is inexpensive but has easier movement and better points than the inexpensive "student" grade pair.

Proportional Dividers.—A pair of proportional dividers (Fig. 4) is a useful tool for transferring exact measurements or scaling up or down. The pivot point can be positioned so that the opposing ends open at a specified proportion to one another. When a distance is measured on one end, the other end enlarges or reduces that measurement proportionally to whatever the setting is.

Enlarging and Reducing Tools

The need to enlarge and reduce in the process of illustrating often arises and there are many ways of going about it.

Graph Paper.—Graph paper of two sizes can be used. For enlarging, place the object on a small grid and block out the outlines on a larger grid. For

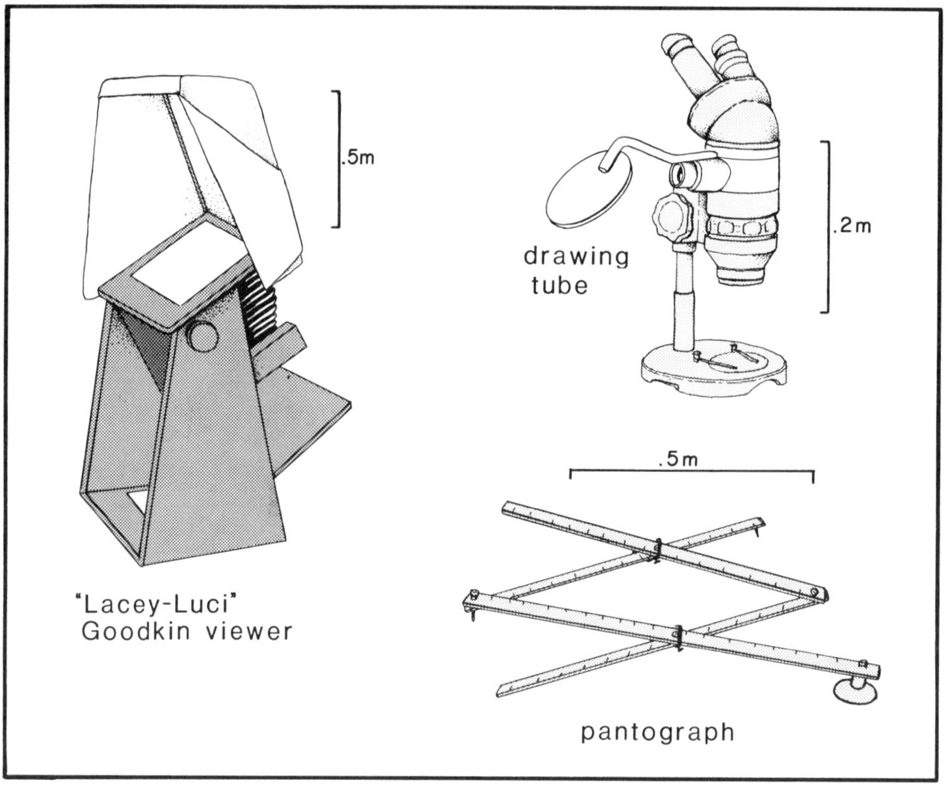

Fig. 5. Lacey Luci, drawing tube, and pantograph.

microscope work a micrometer disc with a grid etched into it can be purchased.

Dividers and Proportional Dividers.—See discussion above.

Pantograph.—A pantograph consists of four perforated strips or bars that are connected by pins forming a jointed parallelogram (Fig. 5). With specified attachment points a drawing is followed with a tracer point and is automatically drawn at the desired size by a pencil point. It can be used to enlarge or reduce from six to 25 ratios depending on its size, the attachment of pins, and position of tracer and pencil points.

Slide Projectors.—A slide projector is useful if you work from slides. Desktop direct-viewing screen models are now available. One made by Tim Simon, Inc. enlarges slides five, eight, and 22 times on an eight by eight inch screen.

Opaque Projectors.—Opaque projectors are available in a variety of forms and range in cost from about $20 to $1500.

We use a Goodkin viewer (Fig. 5) that projects an image ranging from one-fifth to five times the size of the original. The subject is projected through an adjustable lens system onto a ground-glass copy board. The image is traced onto tracing paper and then must be re-transferred onto bristol board. Some distortion occurs near the margins. This type of projector is sometimes called a "Luci Machine" or "Lacey Luci." The more expensive models project the image directly onto the bristol board.

Camera Lucida.—A camera lucida is a lens and prism mechanism that transfers a microscope image onto paper or a wall. Usually the room must be darkened when using a camera lucida.

Drawing Tube.—For compound microscope work a drawing tube (Fig. 5) for two-dimensional subjects, such as leaves of mosses or chromosomes, is commonly used. A lens and prism directs an image onto paper with the artist's hand and pencil point superimposed so the artist can watch the pencil outline the image. The drawing tube is less effective with three-dimensional objects. When there is some relief in your subject, you should carefully check the subject directly to catch any distortions.

Photocopiers.—A photocopier (such as a Xerox copying machine) with reducing and enlarging capabilities can be useful in changing the size of a sketch to better suit your needs. It is especially useful in preparing a base map tailored to the image area you need to show. You can also use it to determine what reduction will do to your choice of lettering size, amount of detail, etc. Several companies now manufacture a zoom model with a continuous range of reductions and enlargements from 50 to 150 percent.

Photocopying is to be used as an aid for preparing artwork and for review copies of line art. Photocopies are not suitable for submission to publisher or printer as artwork for reproduction.

Reducing Lens.—Some artists use a reducing lens to determine how their artwork will appear after reduction. It is the opposite of a magnifying glass, with a concave lens that reduces the image of an object. Observing the illustration from a distance can reveal the effects also.

Lettering Guides

Lettering Set.—A lettering set includes a scriber and assorted lettering templates and pens (Figs. 3 and 6). K&E's Leroy is perhaps the best known, but Staedtler and Faber-Castell also make quality sets. This is the best way to label an illustration that is expected to last for years because it uses the same drawing ink used on the rest of the plate. Figures 10, 11, 22, and 24 show examples of lettering using a Leroy set.

Although lettering sets are initially expensive, the investment is a one-time-only cost. Once purchased the set is always accessible and adaptable to any plate. The style is limited but letter sizes and line weights are easily changed with the different templates and pen nibs. An adjustment of the scriber enables it to make slanted letters (italics).

Correct positioning takes some planning, and practice is necessary to achieve correct looking letter- and word-spacing. Mistakes can be corrected with an electric eraser and shield. The novice can letter on a separate sheet of paper and paste it on the plate.

The lettering set can be used in making dots on a distribution map. The letters "O," "A," "V," and "M" can be used in outlining circles, triangles, and squares. The "×" and "+" can be put to good use as well.

Kroy Lettering.—The Kroy lettering machine (Fig. 6) provides a simple meth-

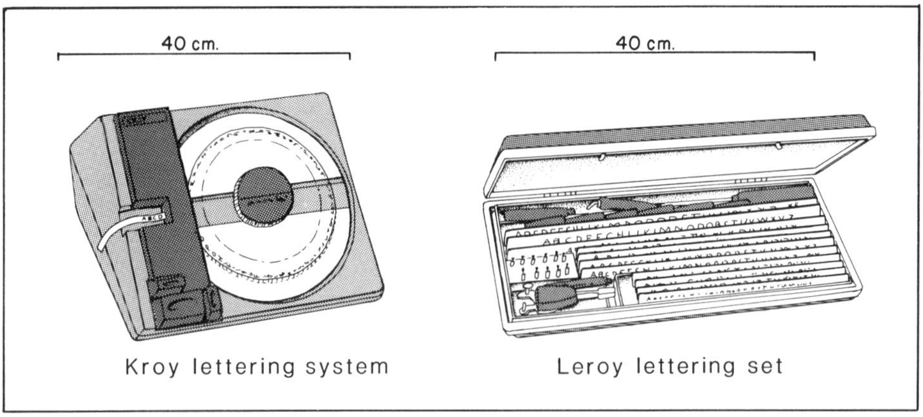

Fig. 6. Kroy lettering machine and Leroy lettering set.

od of producing professional-looking lettering. The operator dials a revolving disk to the desired letter, presses a button, and the letter is printed on an adhesive-backed tape with a protective backing. The tape can be directly affixed to the artwork. Proper letter alignment and word-spacing is automatically set. Proportional letter-spacing can be easily adjusted.

Type disks are available from Kroy in sizes ranging from 8- to 36-point type and in many different type styles (see section on figure labels). The tape cartridge, with adhesive-backed tape and carbon ribbon, is easily loaded into the machine. Cartridges are available with black carbon for black letters on either clear or opaque white tape or white ribbons for white letters on clear or opaque black tape.

Type size and placement should be carefully planned before the labels are made. Light blue pencil guidelines should be drawn for proper alignment. For illustrations with extensive lettering you may want to print all the lettering at one time to determine the best placement. The lettering can be removed from its backing and temporarily affixed to a glossy surface or a silicon-treated carrier paper such as the backing sheet for press-type. A pair of tweezers is handy for transferring and moving the tape strips around until satisfactorily aligned. Long strips can bend and flex as they are pulled into place, so they should be cut into short sections. A gentle burnishing with a protective cover sheet secures the tape. (The word burnish is used as a verb in art instructions meaning to rub with a tool.)

The carbon image on the plastic tape can inadvertently be smeared or rubbed off and must be either sprayed with a fixative (protective coating) or laminated with Scotch tape or a protective tape made by Kroy.

Stencil Lettering Guides.—Plastic stencils can be used for labeling plates quickly and inexpensively. Rapidoguide (Fig. 3) and Wrico are two available brands. The templates come in various sizes designed for corresponding sized technical pens.

Typesetting.—Another possibility for labels is to have them typeset by a professional graphics establishment, but this may be expensive and may require some expertise for placing an order. Typesetting is recommended whenever there is a lot of labeling on a plate. The copy to be cut and pasted can be given a coat of wax on the back for adhesion. The typesetter will usually wax it for you.

Dry Transfer Letters.—A common method of lettering illustrations is the use of dry transfer letters (also called press-on or rub-on letters). Some of the brand names are Letraset, Charrette, Mecanorma Normatype, and Formatt, each with a large range of letters, numbers, symbols, flourishes, and borders in different styles and sizes. (For convenience of this discussion we refer to them all as "letters.") Type faces and point sizes are not consistent between brands even if called by the same style name, so be sure to use the same brand of transfer letters through any particular lettering job.

Transfer letters are printed on the bottom side of a transparent acetate sheet with a layer of a special pressure-sensitive adhesive material on the outer (lower) surface of the printed image. This is protected from sticking to unintended surfaces and dirt particles by a silicon-treated protective backing sheet. The backing sheet can be used in the burnishing process described below and should be saved for storing the letter sheet.

To transfer letters to the artwork, align the letter sheet over guidelines drawn on the artwork and rub over the entire image to be transferred with a burnishing tool. Your guidelines should be either in light non-reproducing blue or placed far enough away from the letter where they can be erased without destroying the letters. As the letter is transferred it appears to turn gray. Carefully lift the sheet, making sure the whole letter adheres to the paper. A final burnishing is necessary to firmly affix the letters. Place a backing sheet or a piece of tracing paper over the letters and burnish firmly. If necessary, corrections, revisions, and repositionings can still be easily made by lifting the letters off with eraser, razor blade, or transparent tape. With the latter, once the letters are repositioned, trim the excess tape away and burnish into place.

For artwork that may be handled a lot, apply a spray fixative to coat the transferred letters with a protective non-glare clear layer. Fixative sprays, such as Letraset Matte or Krylon, are available in aerosol cans. Such sprays should be used sparingly and avoided when possible as they may stain and yellow the paper in time.

Letraset makes a word positioning kit to be used with transfer letters. The letters are rubbed onto a special sheet of paper, then transferred to a transparent film that can be easily positioned, and finally transferred to the artwork.

Letters printed on adhesive-backed acetate are also available for the artist to cut and position. The imprint on acetate is more durable than press-type, but they can be more difficult to align. They are useful for pasting labels on glossy surfaces such as arrows on tone screens on maps.

When lettering a photograph with press-on letters, the letters must be rubbed gently so as not to groove the soft emulsion surface. The letters can be transferred more gently by using the Letraset word positioning kit mentioned above.

When press-on letters are used, the artwork should be mounted on a fairly rigid mounting board, such as 16-ply poster board. The artwork should also be protected with an overlay sheet to which the letters will not stick, such as tracing vellum. If the illustration is to be dry-mounted to a stiff board by a "hot" process (see section on bonding), do this before using the self-adhesive lettering or tone screens, because the heat used in bonding may weaken the adhesive substance. Without these precautionary measures, letters may pop off or be rubbed off in handling.

A precautionary measure, in case some of the letters are lost in handling, is to make a photocopy of the original. You can further allay any fears of damage or loss of letters on the original by having a quality photostat made of your finished work to be submitted in its place (see section on mounting plates). Copies will be needed for reviewers also.

Press-on letters have a limited shelf life, which varies depending on the brand and storage conditions. Unused letters should be stored in a cool, dry place with their protective backing sheet. Old letters lose their adhesiveness and crack and peel as they are transferred or will not release at all. Sheets should be checked before purchasing; the back should feel tacky to the touch.

Burnishers for transfer letters come in a variety of shapes and sizes. The point or ball tip burnishers are best for transferring letters and the flat, chisel-shaped ones are best for burnishing larger pieces such as shading films. Experiment with burnishers; if too large or too small they may cause the letters to crack. Ball point pens or pencils will work as burnishers for letters, but will make the sheet messy.

Symbols.—Circles, squares, triangles, arrows, etc. are available in both transfer sheets or self-adhesive cutouts. As with all line art, the symbols must be chosen with the appropriate reduction in mind (Fig. 7). Open symbols, such as O, △, □, should be avoided if much reduction is intended, as they could fill in when printed. It is better to use different shapes if open and closed are to be mixed. Some symbols such as arrows are available in black with a white outline for highlights across map borders and halftones. Self-adhesive white labels (circles, squares, and rectangles) can be used as background for labels on halftone art. Squares can also be cut from white tape.

Tone Screens and Tapes

Patterned screens are used to differentiate areas in maps, graphs, and sometimes plant illustrations. A tone screen consists of a black pattern printed on a transparent film with a pressure-sensitive adhesive on the back side. Unlike pressure-sensitive lettering, the tone screen is cut and peeled off its backing sheet rather than rubbed directly onto the artwork. The adhesive is formulated so that the film can be lifted and repositioned, but holds firmly when burnished. Dot and line screens are available in intensity percentages ranging from 10 to 80 in increments of 10, where zero percent would be the white of the paper and 100 percent would be pure black. Within this range, the screen is further varied by the fineness or coarseness of the pattern. The screens are printed in patterns from 27.5 to 100 dots or lines per inch.

Light tones, those in the range of 10 to 30 percent, may drop out when an illustration is reduced, likewise, dark patterns may fill in. Line patterns should be selected so that after reduction they would be no finer than 50 lines per inch and dot screens no finer than about 100 per inch. As an example, a screen of 32.5 lines per inch will reduce to 65 per inch at 50 percent reduction (see Fig. 7).

Sharp-pointed knives with pencil-like handles, such as X-Acto, are used in cutting and trimming the film. First, cut out a section of film slightly larger than the area to be shaded and separate it from the backing sheet. Lightly burnish the section of film over the area to hold it in place, trim around the outline of the area and remove the surplus film. Then firmly burnish into place, using a flat burnisher with a chisel end.

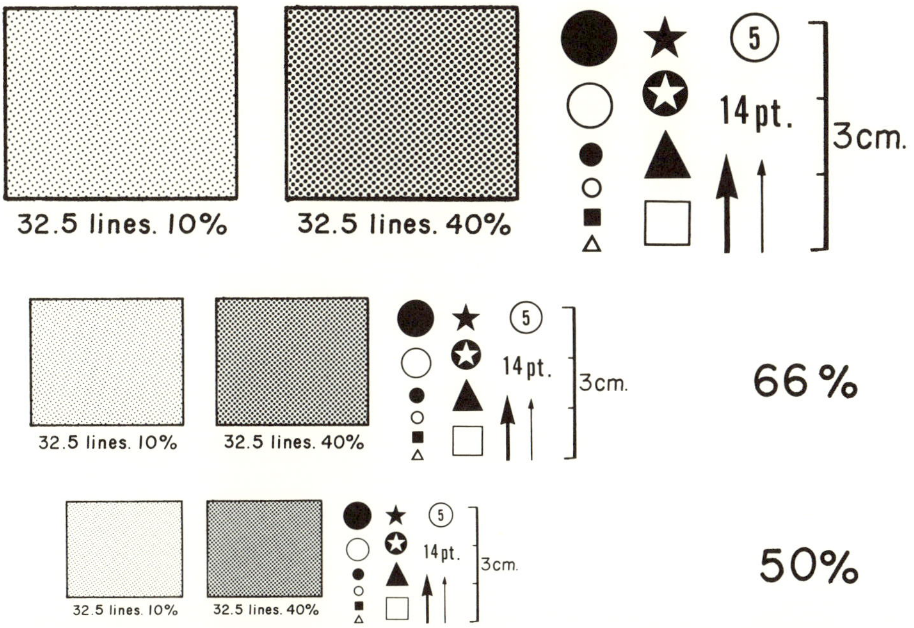

Fig. 7. Tone screen and symbols; upper original size, middle reduced to 66 percent of original and lower reduced to 50 percent.

If the film is burnished and then removed it may take up some of the underlying ink, so you cannot change your mind too often. If it is picking up underlying ink too easily, even without burnishing, the artwork can be sprayed with a fixative before burnishing. Pencil marks, dirt, eraser crumbs, and hair will show through the pattern, so make sure the area is clean before the acetate is laid down.

If the toned area is to be outlined, the outlining should be done before the pattern is applied. If outlines are added later, the ink could bleed under the acetate and make a mess. Touch-up of lines should be done carefully with a fine technical pen.

In using tone patterns to differentiate areas, let adjacent values contrast at least 30 percent.

Screen tones were designed to save the artist time, but the option of hand-drawn patterns is still available. When carefully done, the effect is identical and saves the worry over impermanence. With a free-flowing technical pen nearly perfect dots can be made with even pressure. Ruled lines can be made with a T-square, varying the angle with a triangle. However, sloppy dots or lines look unprofessional so the unsure hand should stick with pre-printed screen tones.

Line Tape.—Press-on line tapes, such as Letraline Graphic Art Tapes, give a uniform width for borders, graph axes, etc. They come in a large range of widths and are sold in rolls with an easy-to-handle dispenser that can be run along a straightedge. Different patterns are available and there is a selection of flexible tapes (Letraline Flex-A-Tape) for graph curves.

Tape

Artist Tape.—3M 285 White Artist Tape is a good tape for mounting drawings to mounting boards and holding drawings to the drawing board or light box. It is white with a low gloss and can be easily written on. It adheres to plastics as well as to paper, can be removed without leaving gummy residue, and does not damage the paper surface. After a year or more the tape may dry out and leave a residue or tear up the paper.

Transparent Tape.—Transparent (Scotch) tape is useful for covering press-on letters. These transparent tapes should not be used for temporary holding tasks on artwork because removal may take some of the surface with it and they definitely should not be used for permanent holding tasks. Scotch makes a repositionable tape (3M 281) that can be removed from most surfaces without damage.

Double-coated transparent tape is useful for mounting artwork and labels. Scotch 3M 666 is easy to use and is supposedly slow-aging.

Packaging Tape.—Several different types of tape are useful and suitable for packaging, but have no place directly on artwork. Masking or drafting tape should not be used on the outside of parcel packages—the U.S. Post Office will not accept it. The adhesion is weak so that it easily becomes unfastened and sticks to other objects. Paper packaging tape is often used, but unless you have a dispenser with a moistener, it can be messy. Plastic package sealing tape is relatively burst- and split-resistant. Filament tape is excellent for reinforcing packages. It is reinforced with high tensile strength glass filaments.

Bonding

Mounting artwork or bonding pieces together can be done in a variety of ways. The selection of adhesive materials should depend on the type and value of the artwork. Any adhesive materials used on a valuable drawing should be removable at a later date. Unstable bonding materials should be used only on non-original artwork (such as photostats), or should be cleaned off after it is returned from the printer. Water-base glue should not be used in mounting halftone art because the resultant wrinkles tend to show as shadows in reproduction.

Rubber Cement.—Rubber cement provides a good bond and is easy to clean up around the edges, but with time it is very destructive. It should only be used for very ephemeral paste-ups, never on anything worth saving. Rubber cement can be easily removed if done within a few months, but within a year noticeable discoloring and curling may occur. A ball of dried rubber cement or a natural rubber block can be used to clean up excess.

For a strong bond apply an even coat of cement to both surfaces, let them dry and join them with firm pressure. For large-scale work you can buy special dispensers with an adjustable-length brush mounted in the cap for proper depth of dipping.

Rubber cement thinner, such as Bestine, can be used to maintain the cement at a good flowing consistency, as well as to clean up and remove rubber cement. A Valvespout thinner dispenser with a closable spout is handy for small jobs.

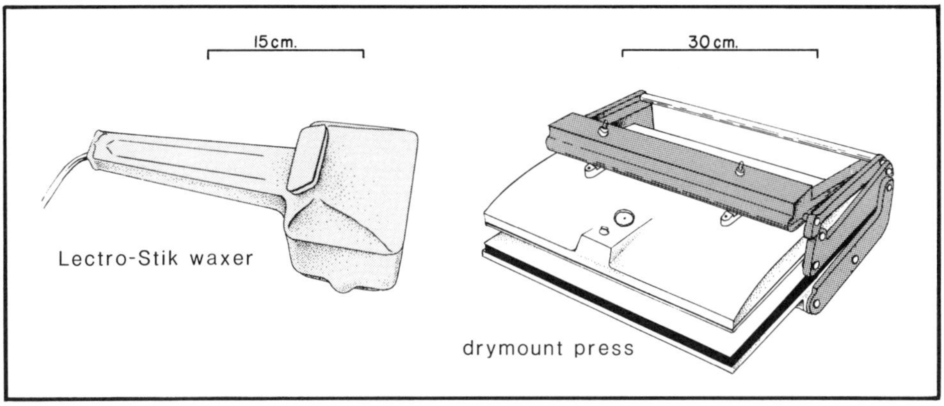

Fig. 8. Lectro-Stik Waxer and dry mount press.

Spray Adhesives.—Spray adhesives come in aerosol cans. Although no longer containing chlorofluorocarbon propellents, they are still unhealthy to breath. Like rubber cement, they are damaging to paper over time, but, unlike rubber cement, they are not removable.

White Glue.—A water-base glue such as Elmer's is excellent for porous surfaces such as paper. It should not be used in mounting photos or on surfaces to be heated. When using it for bonding paper, spread thinly to prevent warping. White glue, technically known as PVA (polyvinyl acetate), is the safest glue for artwork to be stored permanently mounted, but it cannot be easily undone, so the backing board should be as acid free as the drawing board.

Glue Sticks.—Glue sticks are recommended for photos and small paste-up jobs. The adhesive substance is a wax-like material that comes in a lipstick-type container and is easy to apply. Pritt Glue Stick, Carter's Stix-a-lot, and Peli-Fix are good.

Waxing.—The Lectro-Stik waxer (Fig. 8) is useful for paste-ups. It is a hand-held tool with a heating element that melts the wax within an enclosed reservoir. The molten wax coats a roller that transfers a uniform layer of wax to the back of the piece to be mounted. The thin layer of wax dries almost instantly and will not stick until burnished. The coated material is easy to position and can be peeled off for repositioning. When the artwork is positioned, place a protective sheet over it and rub with a flat-edged burnisher.

If necessary, the wax can be removed by scraping with a single-edged razor blade or dissolved with rubber cement thinner.

More expensive wax coaters are available in which the artwork is fed directly through rollers and is evenly coated on the backside. This is the system used in professional print shops.

Dry Mounting.—Dry mounting provides a relatively permanent mount in that artwork cannot be repositioned. A heat-sensitive or pressure-sensitive tissue provides an instant and even bond. Artwork should be carefully laid out before bonding.

The dry mount press (Fig. 8) is an expensive item, but affords a superior method of bonding materials such as photographs to a mounting board. The press consists of a heavy, flat, thermostatically heated, metal platen that is pressed evenly against the mount. A small hand-held tacking iron is used for positioning the mounting tissue on the back of the artwork and then onto the board so that it will not slip out of position when placed in the press. Also, a heavy, flat, metal weight is used for cooling and pressing flat the freshly bonded artwork.

The materials for dry mounting include release paper and the mounting tissue, which is available in sheets and rolls. Release paper or "carrier" is a silicon-treated cover sheet used to protect the platen and the artwork. A mounting tissue that works well for the purposes described in this article is Seal MT5. It is a waxy-feeling paper that is placed between the two pieces to be bonded and they are pressed together at a temperature of about 225°F for about 45 seconds or more depending on the thickness of the board.

If the press seems too costly for the amount of anticipated use, a household clothes iron can be substituted. The mounting tissue will include instructions for temperatures and duration of heat application. Before bonding you should pre-dry the artwork and mounting board with the iron at a low setting to prevent moisture bubbles from forming when bonding. Tack the seal tissue to the back of the artwork and trim it to size, tack the work to the mount board at one side or in the middle and slowly iron it, working toward the free edges.

If the illustration is to be dry mounted onto a stiff board, such self-adhesive applications as press-on letters and tone screens should not be applied until after the mounting, because the heat used in bonding may weaken the adhesive substance.

A simple form of dry mount is the "cold" mount using a double-faced adhesive material. You can buy sheets or tape with a protective backing or a mount board with the adhesive already on it. Cold mounts are useful if you must avoid the use of heat or wet adhesives.

In cold mounting, to avoid the formation of air bubbles between the print and adhesive surface, apply it from the top edge and roll it down, pressing as you go. If air bubbles still form, prick them with a pin and flatten them out.

Cutting Tools

The comfortable plastic-handled Fiskars scissors with stainless steel blades keeps a good edge and is a pleasure to use. A small pair of scissors is handy for cutting tape.

The indispensable (but disposable) single-edge razor blade is inexpensive, especially when bought in quantity. It is useful for microscope work as well as trimming paper.

Pencil-handled cutting knives come in different sizes with replaceable blades. The X-Acto knife (Fig. 2) can be used for delicate cutting. A swivel knife has a blade which freely rotates. This knife is especially designed for cutting curves, but it takes some practice before full control is gained. A large knife (X-Acto #5) is useful for cutting heavy board.

Although a paper cutter is not essential, wherever one is available it gets heavy use. It should be checked for correctness of square cut and sharpened periodically. There is the conventional arm-blade type and the more precise rolling-disk blade type, such as the one made by Honeywell-Nikor ("Safety Phototrimmer"). This latter type can be found in some photographic supply stores.

Do not forget the first-aid kit.

Cleaning

A clean, soft dusting brush, not the side of the hand, should be used for brushing eraser crumbs away.

A water jar, a pair of tweezers, and tissue paper are useful for cleaning inking tools. A little ammonia and liquid detergent added to the water will cut stubborn dried ink. A water tap and sink are useful if nearby.

"Wet Ones" (moist towelettes) are handy for cleaning hands and drawing surface, and should be used frequently to avoid fingerprints.

A chamois or soft cotton cloth is useful for wiping ink off pen nibs.

Rubber cement thinner is a good solvent and cleaner. It will remove most oily residues that repel ink on bristol board and drafting films. However, it is highly volatile and flammable, especially in dry winter air loaded with static electricity.

Rubber cement itself makes a good cleaner for oily spots on paper. Brush the rubber cement on, let dry, and rub off with an accumulated ball of dried rubber cement.

Size and Proportion of Plate

From the editor's and printer's standpoint the best illustrations are those designed to fit the journal to which they are submitted. Check the journal's instructions for any specifications in size and preferred reduction, or check with the editor before taking any extraordinary liberties. A safe format for the artist to work in, if the page size is unknown, is a plate size one-and-one-half times as high as wide.

If a journal page has two or three columns, the options may be open for illustrations designed to fit the width of one or two of them. Some illustrations, such as a map requiring much horizontal detail, can be made in two halves for placement on facing pages. When designing such an illustration, the matching margins of the two facing plates should have an obvious breaking point so that no information is lost if exact alignment does not occur. Try to avoid wide illustrations intended to be printed sideways on the page. Although accepted practice, it runs across the grain of the reading pattern.

Reduction of line art for printing generally improves the appearance. Reduction seems to lessen flaws; it smooths out unsure lines and evens out some tones. You should be aware that a very fine line may disappear and an excessively busy or heavily stippled area may fuse or darken. What reduction does is make big flaws into little flaws; it does not eliminate them. The actual improvement comes from working at a larger scale where manual dexterity and visual alignment are less restricting.

The amount of reduction has to be taken into consideration when choosing line thickness, spacing, letter size, and density of stippling. Most well prepared line drawings can be reduced to 50 percent and should be drawn with that intent. Photographs or any other halftone art, on the other hand, usually lose some of their quality when reduced. Ideally, halftone art, should be reproduced at the same size, but if this is not possible it should be reduced no smaller than 66 percent of the original. Any consideration of enlarging artwork should be dismissed out of hand. Minor flaws in line drawings or the graininess in photographs would be emphasized by this.

It is important to understand the difference in reduction instructions. To reduce to 50 percent or by one-half means the original is two times the size of the printed (or reduced) size and to reduce to 66.7 percent means the original is one-and-one-half times the intended reduced size. A circular proportional slide rule (Fig. 3) is available for calculating percentages of original size and the number of times of reduction, or the inverse. Alternatively, these figures are easily obtained on a pocket calculator. Divide the planned reduced width by the width of the original, and the resulting fraction times 100 is the percentage.

Plate numbers and instructions to the editor are best put on the back of the illustration or mounting board, which the editor can transfer to the front of the plate outside the image. When assigning scales in the figure legend be sure to take into account the reduction factor. If this is neglected and is not caught by the editor, it will result in either expensive line changes in proof or published errors. Scale bars in the illustration itself are preferable (see section on figure labels).

Because these instructions are aimed primarily at plant taxonomists preparing illustrations for publication in botanical journals, a recommended maximum size for illustrations (including the mounting board) is that of a standard American herbarium sheet, 11½ by 16½ inches (30 × 42 cm). This is a convenient size for temporary storage by the editor (often a taxonomist with appropriately sized genus covers for protection and cubbyholes in a herbarium case), for shipment in herbarium specimen cartons, and for possible permanent deposit in the herbarium after it has been reproduced by the printer. This means that a full-page plate for a five by eight inch journal page prepared for 50 percent reduction will have a minimal margin.

Ideally, an illustration should not measure the full page height, allowing enough space for the printed figure caption. When the figure caption has to be printed on the opposite (facing) page, it introduces problems in page make-up of the article. To calculate caption size for the printed figure, make estimates from published examples in the intended journal. In this publication, for example, each line of the caption is ⅛ inch (3 mm) high and there is about ⅛ inch space between the lowermost margin of the illustration and the first line of print. When working with 50 percent reduction in mind, remember to double the size of the letters.

At least one inch margin above or below the image is preferred by the editor and printer to place reduction and orientation instructions and plate identification where it can be included in the printer's photograph.

If there is less than a full plate of information, try to accommodate the full width and keep height to a minimum. A well designed half plate looks better than a sparse full plate, and costs less.

Figure Labels and Captions

Labels

Most illustrations in botanical articles have labels (letters or numbers) that identify the parts, and scale indications. Lettering and scales should complement the illustration by being neither too bold nor too delicate in relation to the drawn lines. Hand lettering is not recommended unless it is done by a well-practiced artist with a steady and even hand. Typewritten labels are unacceptable to most journal editors. A mechanical lettering device or press-on letters should be used.

Labels should be carefully placed to obviously correspond to the indicated figure. When possible, the units in an illustration of a plant should follow a logical progression around the page (clockwise is common). The progression should follow the sequence used in a description, such as habit, stem, leaves, inflorescence, flowers, and fruits.

Attractive, easy to read typeface styles are Baskerville, Folio, Franklin, Futura, Gothic, Grotesque, Helvetica, and Univers. They are available in a variety of forms: typeset, press-on, stencil, and Kroy. These typefaces come in roman and italic, and include caps (capital letters), small caps, and lower case, in light, medium, and bold weights (line thickness) (Fig. 9). Avoid using a mixture of styles (Helvetica, Folio, etc.) on the same plate or even on a series of plates making up an article. The use of bold, roman, and italic type in the same style can provide variety if necessary.

Typeface styles fall into two main categories, serif and sans serif. *Serifs* are the short strokes that project from and at an angle to the ends of the main strokes of the letter. Baskerville (Fig. 9) is a serif typeface. *Sans serif,* as the name implies, are letters without serifs, such as Helvetica (Fig. 9).

One-weight sans serif, in which all lines are the same weight, may be more desirable for line art because of its complementary simplicity. Lower-case letters are easier to read than caps because of their more familiar and more distinctive appearance.

Printers have their own units for measuring type size and their system is used in most of the various forms of lettering you will select. Type size is measured in *points.* There are 72 points to an inch. The length of a line is measured in *picas.* There are approximately six picas to an inch.

The type you are now reading is 10-point Optima, a sans serif style. In this text an extra point space called linespace or leading (pronounced ledding, as in *plumbum*) has been added between each line, so a measurement from baseline to baseline is 11 points. Eight-point Optima with a one-point linespace is used in the figure captions and literature-cited section. The block of text on this page is 30 picas wide and 48 picas high.

Helvetica condensed - Hemichaena fruticosa
Helvetica light - Hemichaena fruticosa
Helvetica bold - Hemichaena fruticosa
Baskerville italic - Hemichaena fruticosa
Baskerville light - Hemichaena fruticosa
Baskerville bold - Hemichaena fruticosa

Helvetica condensed - Hemichaena fruticosa
Helvetica light - Hemichaena fruticosa
Helvetica bold - Hemichaena fruticosa

Helvetica condensed - Hemichaena fruticosa
Helvetica light - Hemichaena fruticosa
Helvetica bold - Hemichaena fruticosa

Fig. 9. Samples of Helvetica and Baskerville typeface styles. The 18-point Helvetica is reduced to 12-point type (or what the upper would look like reduced 66%) and 9-point type (or what the upper would look like reduced 50%).

Letters vary in width, the i and l being narrowest and the W widest. Consequently they require proportional spacing. The standard typewriter gives equal spacing to all letters; 10 pitch ("pica") is 10 characters to the inch and 12 pitch ("elite") is 12 to the inch, no matter what the letters are. It is a poor match in appearance to the typeset text used in most journals. Some special typewriters and word processing printers (and software) offer proportional spacing.

Correct letter-spacing is not always simple. Character combinations of letters within words, such as those with the capital letters A, L, T, V, and Y, appear too far apart in relationship to other letters and have to be set closer together to appear correctly spaced to the eye. The printer refers to this adjustment of space between two letters as "kerning." The letters in the two generic names on the top line below are equally spaced but appear irregular. The illusion of equal spacing is given to the names on the bottom line by selectively reducing the spaces between the first three letters of each name.

LYCIUM TAVERNIERA

LYCIUM TAVERNIERA

When hand-spacing, a good reference is to leave the space between words that the letter *n* uses within a word. Allow two *n*'s between sentences.

Superscript and subscript letters and numbers are offset above or below the line and are in a smaller point size. For lettering mathematical equations and chemical formulas see Allen (1977). For more information on typography see Craig (1978).

The size of lettering must be based on the anticipated reduction. Lettering planned for six-point type size on final reduction is the minimum; anything smaller cannot be read comfortably. If an illustration is drawn for 66 percent reduction the type size should be drawn half again the anticipated size, and for 50 percent reduction, twice the anticipated size (see Fig. 9). If more than one size is used in a plate, the middle size should be about the same as that used in the text. An example is the hierarchy of political units on a map, such as country, state, county, city, village, where the county name would be in text-size type.

Labels, scale bars, and arrows should contrast with the background. Problems arise with halftone art, where the labels must be set apart from the middle tones. White or black labels work where the contrast will be significant. Otherwise, a contrasting background must be used. A self-adhesive white circle or square as the field for black lettering can be used. If white letters are preferred but the halftone is not quite dark enough, a square can be cut out of a dense tone screen pattern (60%–80%, 100 line/inch) and used as background. Press-on arrows and letters are available in white, black, and black with a white "shadow" (see Fig. 25).

A scale measurement on the illustration adds to the scientific precision in a paper. For the publication of new taxa, reference to scale is recommended by the International Code of Botanical Nomenclature (Voss, 1983: Recommendation 32D.3). Bar scales are preferable to magnification numbers in the caption because the editor may choose a reduction different from that intended by the author (for examples of bar scales see Figs. 10, 13, and 15). If a plate is reduced by a specific amount, magnification numbers next to parts in the plate work well, especially for quick reference in a field guide.

Captions

The figure caption is the title and descriptive matter below the figure. The word "legend" is restricted in this discussion to the box that appears within some maps and graphs with the title, keys to symbols used, and scales.

An effort should be made to make each illustration as self-explanatory as possible. Ideally the reader should not have to rely on the figure caption and especially not on the text for understanding the figure. Lengthy and redundant figure captions should be avoided.

In the spirit of scientific accuracy, for plant drawings the figure caption should include citation of the actual specimen(s) from which the artist made the drawing, especially for new species. In describing a new species, this will satisfy "Recommendation 32D.2" in the International Code of Botanical Nomenclature (Voss, 1983). However, the illustration may represent a concept of a species rather than an actual specimen, in which case the citation can be left out.

If the scale is not indicated on the illustration by a multiplication figure or bar scale it should be given in the caption. If the illustration is being reprinted or adapted from another source, the caption is a good place to acknowledge it. This is a good place to identify the artist, if it has not been done in an acknowledgment section.

Plant Illustrations

In describing new taxa, authors should, when possible, supply figures with details of structure as an aid to identification.

—Recommendation 32D.1 of the *Code* (Voss, 1983)

A well-drawn illustration is always more effective than a photograph of a herbarium specimen in conveying the diagnostic characteristics as well as the overall likeness or gestalt of a species. It should inform and elucidate beyond what can be described in words. A plant drawing can depict information that the author did not even think of, as well as information that the viewer can easily recognize without having to understand or translate the text. Although a Latin description is required in a protologue to transcend language barriers, the message of an illustration is universally accessible, therefore a drawing of a new species is especially important.

A good botanical artist strives for accuracy in proportions, curves, angles, and shapes, and attempts to create the illusion of three dimensions and surface textures. Although accuracy is more important than aesthetics, the two are not mutually exclusive. An illustration serves as a diagram of what the plant is, not a reproduction of the plant.

Drawings are usually preferable to photographs because they permit the author and artist to either draw a specific individual or present a hypothetical ideal individual that depicts a species rather than a particular plant. Parts from different plants can be combined, important parts emphasized, extraneous parts left out, and hidden parts featured. Certain aspects of a plant can be stylized to convey the maximum amount of information. Artistic license, however, must never lead to misrepresentation.

Although each plant will have its own problems and solutions, certain conventions should be followed. The ideal plant illustration depicts the general habit of the species as well as details of the diagnostic characters that define it at all stages of its life cycle. However, the quality of available material, the need for comparison with related species, the intended use of the illustration, and available funds for its production will determine its scope. The cost of a plate should not influence the amount of information, rather the amount of artistic enhancement should be adjusted. Figures 10, 11, and 12 show three treatments of the same plant. Although Figure 11 shows a minimum of tone and texture, it adequately depicts the species and requires less than half the time for completion that Figure 10 does. The stipples in Figure 10, however, add depth and character and may be worth the added time and cost.

Diagnostic features mentioned in the key should always be drawn, as a minimum. When drawing only one species, you can freely use enlargements that fit the layout of the plate, but if two or more related species are drawn, char-

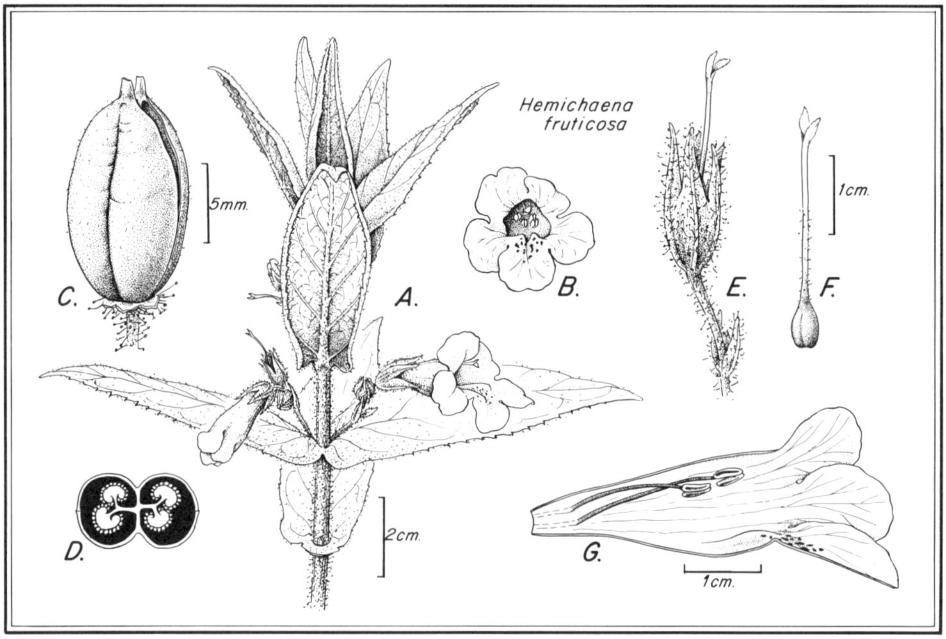

Fig. 10. Illustration of *Hemichaena fruticosa.* Drawn from a herbarium specimen, and the three-dimensional reconstruction with the aid of photographs and pickled material. Drawn on bristol board with Hunt #102 quill pen, the lettering was done with a Leroy lettering set. Drawing time: 3½ hours; cost from the printer for photography and proofs (fine line): $10.00.

acters used to distinguish them should when possible be drawn at the same scale, in the same view, and in the same relative position. The artist and botanist should review the range of material considering the extremes before deciding on the scale of enlargements.

Whenever an enlarged detail is drawn its relationship to the plant should be obvious. The detail can be placed next to, or on top of, its location on the plant (Fig. 13), or can be connected with a line or arrow. The reader should be able to understand what is illustrated without referring to the caption. Enlargements of details should be proportional to the complexity of the details to be shown, as well as proportional to their importance to the species identity.

If pendant flowers or fruits are separated from the habit they should be drawn hanging down. An exception would be a longitudinal section, which may be easier to read if right side up.

It is best to dissect the same flower that is drawn entire, or at least show the dissection as comparable to the original. A longitudinal dissection of the flower is not always necessary. Often a flower is better represented by removing part of the perianth, or by cutting away just enough to expose the inner structure. When plant parts are eliminated, this should be indicated by scars or obviously cut edges (as in *Nemacladus,* Fig. 14).

In drawing leaves it is usually easiest to sketch in the midrib first to establish the general placement and direction, then add the leaf outline and the details of the margin. Opposite leaves should be drawn slightly askew to avoid the problem of direct foreshortening, as well as to show the nodes and stem. If

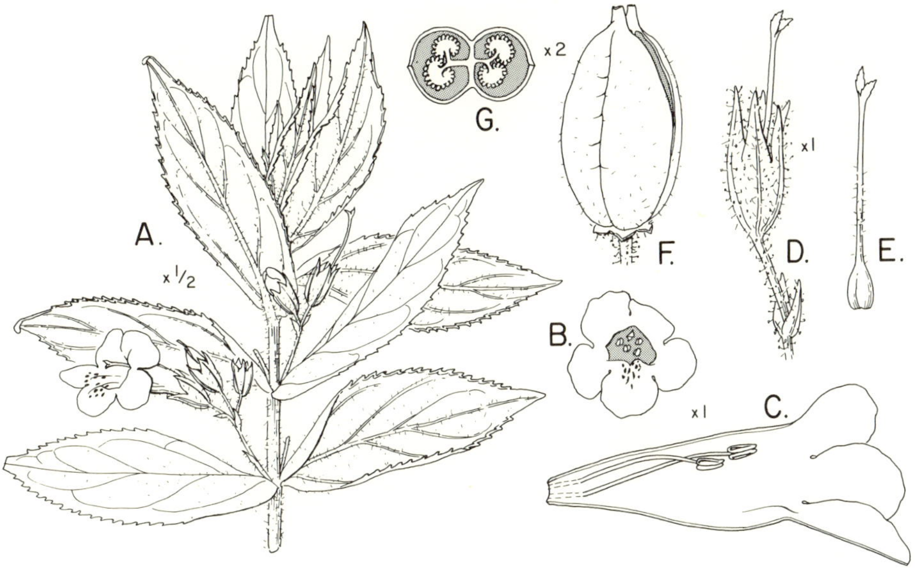

Fig. 11. Illustration of *Hemichaena fruticosa.* Drawn from a herbarium specimen; unreconstructed. Drawn on bristol board with technical pens, 0 and 00, and tone screen, the lettering was done with a Leroy lettering set. Drawing time: 1¼ hours; cost from the printer for photography and proofs (fine line): $10.00.

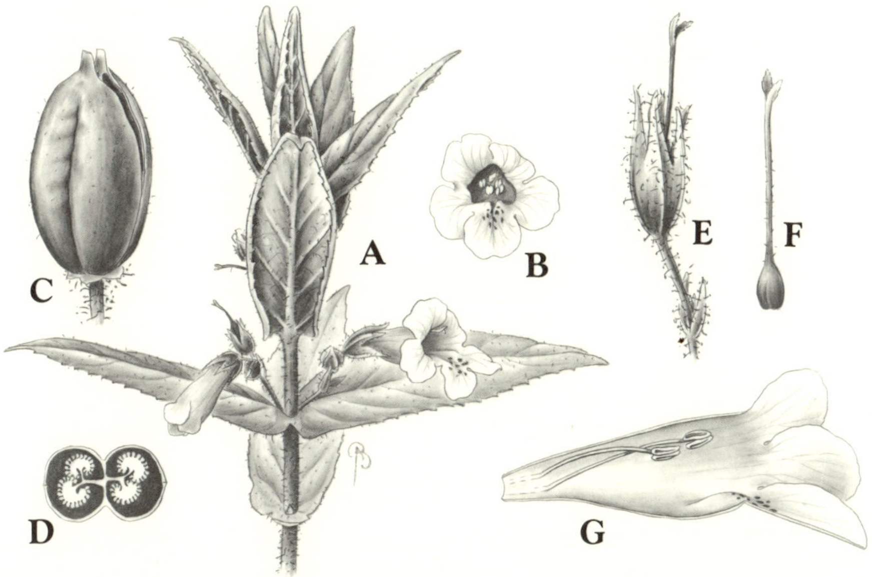

Fig. 12. Illustration of *Hemichaena fruticosa.* Drawn from a herbarium specimen, and the three-dimensional reconstruction with the aid of photographs and pickled material. Drawn on drafting film with H and 5H pencils, the lettering was done with press-on letters. Drawing time: 2 hours; cost from the printer for photography and proofs (screentone process®): $15.00; printing cost more expensive than Figures 10 and 11, may require better paper for reproduction.

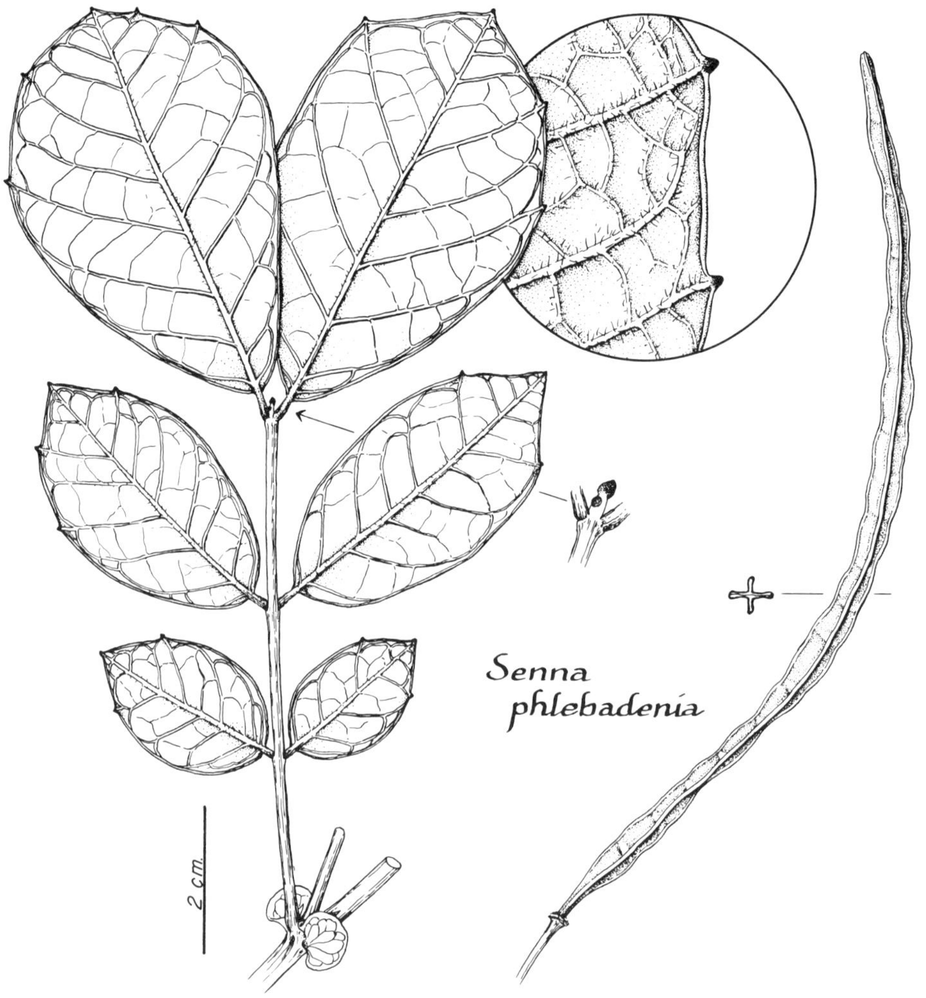

Fig. 13. Illustration of *Senna phlebadenia* showing circled enlargement of leaflet margin, an enlargement of the petiole-appendage, and a cross-section detail of the fruit. From Irwin and Barneby (1985, p. 193), and used with the permission of the authors.

the leaves are crowded, omit one or a few to show their arrangement on the stem.

If the leaf is bent, the midrib must follow the bend. Stems, veins, etc. almost always have a curve; do not use a straight edge. Flowers and leaves must be correctly placed on their axes (see *Asclepias,* Fig. 14).

For some features the front and back sides should be drawn, even if the plant has to be distorted from its natural aspect. If leaves have an unusual pubescence or distinct venation beneath that needs to be shown, flip one or more of the leaves over in the drawing.

In some ways drawing from pressed specimens may be easier than from live plants. The artist does not necessarily have to reconstruct a three-dimensional plant, and it is often better not to when showing features such as leaf size and shape. The artist should, however, correct such distortions as the shrunken

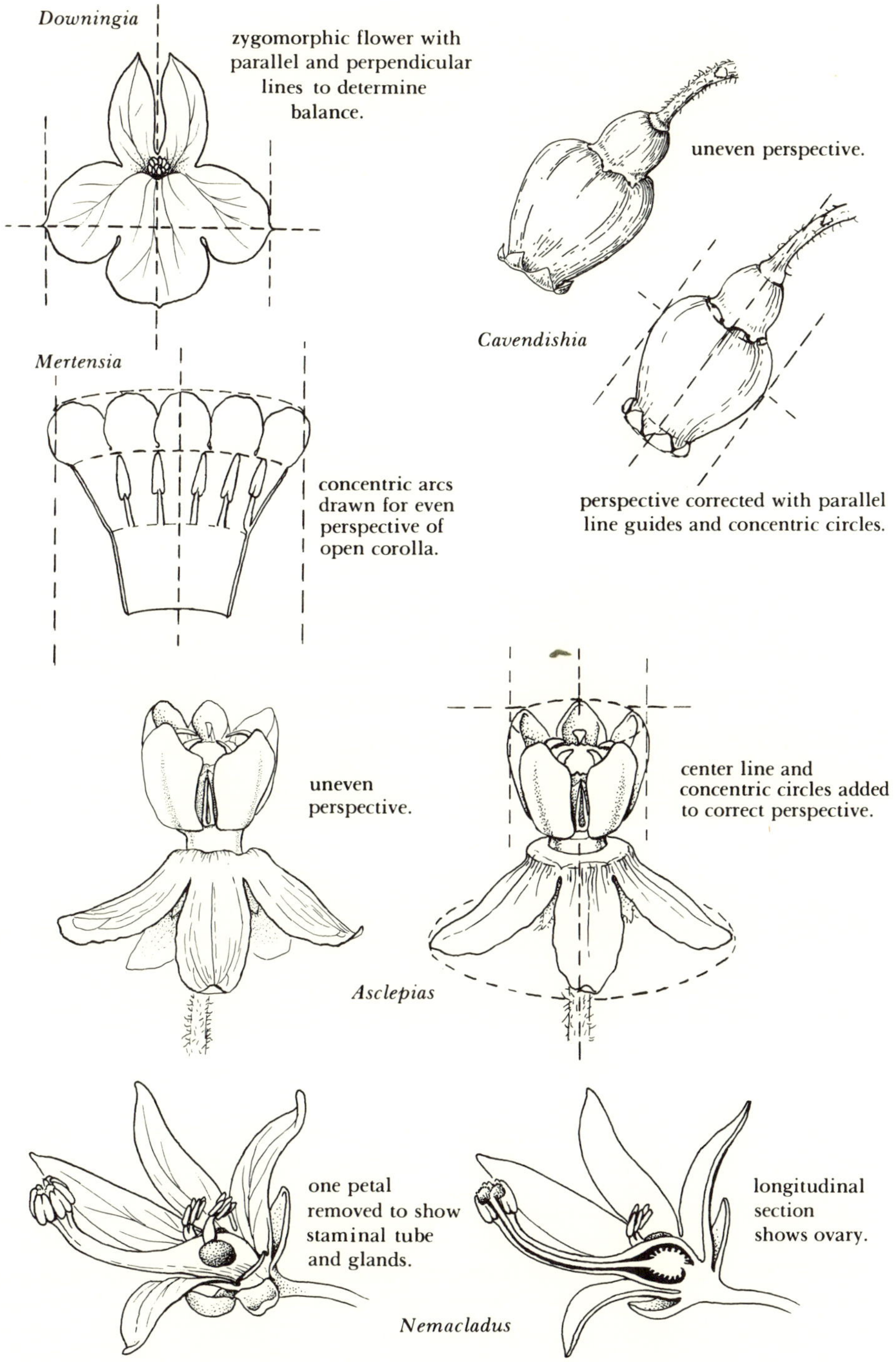

Fig. 14. Flowers showing methods used in drawing them. Explanation given in the text.

aspect of dried fleshy leaves or the flattened dimensions of a tubular corolla. Boiling or soaking flower parts can help correct this if the author cannot provide pickled material or photographs. To some extent the original shape of dried and pressed flowers can be restored by immersing in boiling water for about a minute or soaking in a dilute detergent solution.

To ensure a well-arranged plate, rough sketches can be done on scratch paper and then arranged and transferred onto the final drawing sheet (bristol board). The plate should be designed to make the most effective use of space. Often what started out as a full-page drawing can be rearranged to fit a smaller plate. The plate size, specified by the journal, should be marked out, the sketched parts arranged within the borders, and then traced onto the bristol board over a light box (or directly onto drafting film). Another option is to draw directly onto bristol board, then cut and paste up the plate. The common fault with this method is that it is easy to set the pieces too far apart; the shapes of the cut-out pieces bias your view of the composition so that it is difficult to imagine the plate as it would appear without them. Before the pieces are firmly attached, viewing the plate through a sheet of tracing paper or making a photocopy will enable you to see it without the lines to make sure the pieces are not too spread out or too cramped. Space for labels, scales, and titles should be considered at this time. The plate should have a logical arrangement so that labels can be placed orderly, such as clockwise around the plate corresponding to the order given in the description (habit, leaves, flower, fruit).

With a sketch transferred onto bristol board, the artist can then add details in pencil directly on the bristol. If the artist begins the rough sketch on the bristol, he or she may be less likely to make changes in size and layout, or may end up with a confusing mess of half-erased lines. The final pencil sketch should be carefully examined before any inking is done, making sure that the symmetry and perspective are correct and that everything is carefully aligned within the page borders. Stamens and like parts should lie along even lines, the central axes should run parallel to the edge of the plate.

Perspective is important in producing the illusion of three dimensions and must be considered when reconstructing flowers and fruits. Concentric circle and oval templates are useful for creating correct outlines (Fig. 14, *Asclepias* and *Cavendishia*). Although flowers and fruits are not always symmetrical in nature, it is best to balance them diagrammatically to give a clear image. Zygomorphic flowers (bilateral symmetry) can be made equal on both sides by tracing one half backwards (Fig. 14, *Downingia*). Actinomorphic (regular) flowers can be made symmetrical by working within a circle or square, or, for an angled view, an oval or trapezoid (Fig. 14, *Asclepias*). A sketch can easily be checked for symmetry by turning the drawing upside down or looking at it in a mirror; you will see at once if the eye had distorted the image as you drew.

If the sketches are properly prepared, inking becomes only a matter of a steady hand. The inking proceeds much more smoothly when you are confident that the sketches are accurate and the plate design is well thought out.

In scientific illustrations, the shading convention is with light coming from the upper left. This should be understood during the sketching process and executed during the inking, regardless of whether or not a light was present. The resultant shadows will determine the three-dimensional aspect, relief, and highlights (Fig. 10 vs. Fig. 11, fruits). A simple heavy line on the lower right, or a hint of stippling, goes far to define a subject. The effect of halftones for shadow, form, and texture is achieved with stippling, cross-hatching, or contour lines. Different plants demand different techniques. Consider the differ-

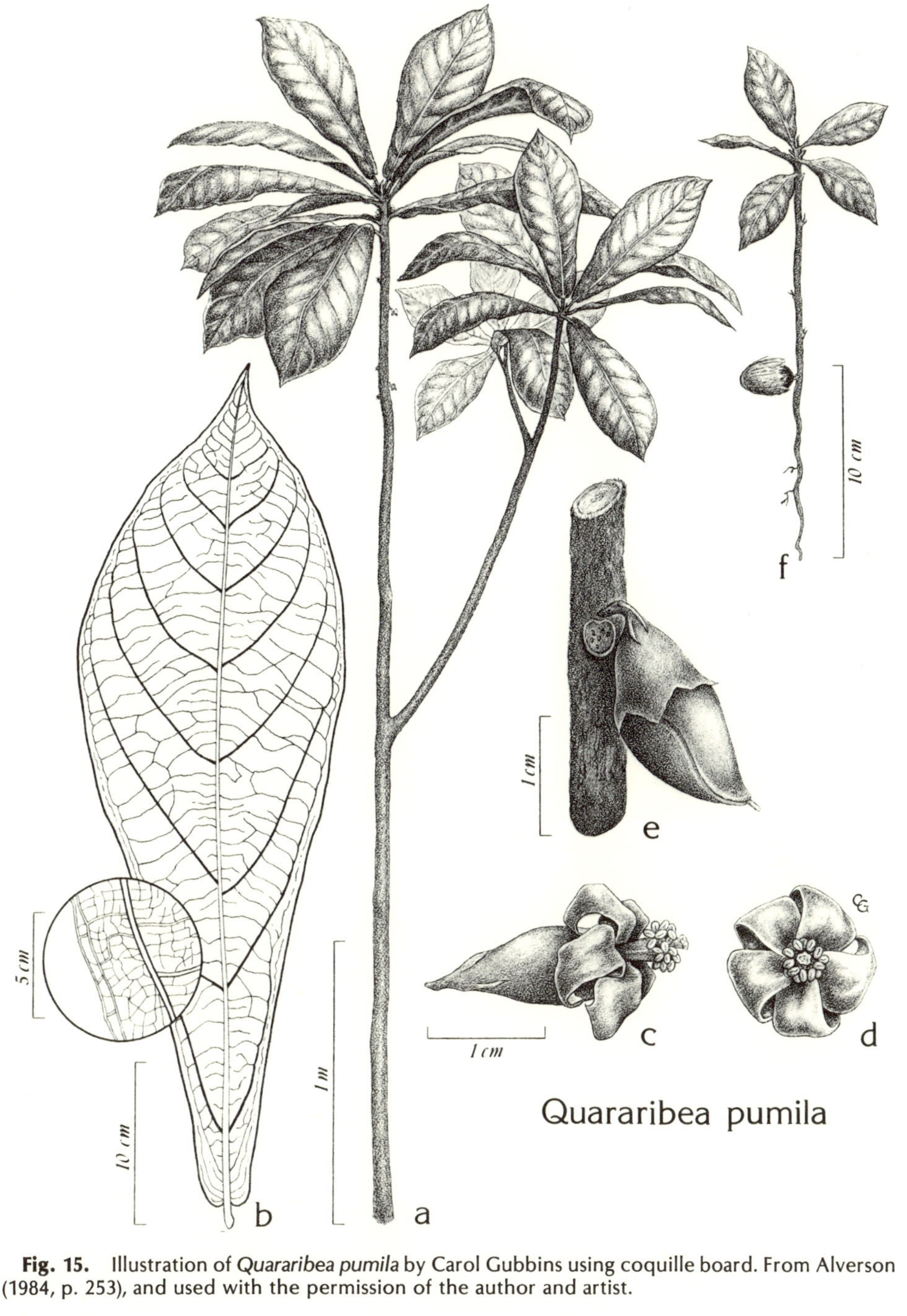

Fig. 15. Illustration of *Quararibea pumila* by Carol Gubbins using coquille board. From Alverson (1984, p. 253), and used with the permission of the author and artist.

ences between monocots and dicots, fleshiness and woodiness, pubescence and glabrosity.

Emphasis of details in line drawing and optically effective gradation of tones should make the drawing more comprehensible. Attention can be called to areas of importance by use of heavier lines or by increased detail. Background objects can be de-emphasized with thin or broken lines or stipple outlines. It is important that the artist understand that line drawings are limited to black and white. Weak lines, such as those drawn faintly to create a background image, may not photograph well.

Most botanical illustrations are drawn in pen and ink because of the ease of printing and legibility. Similar effects are obtained whether the artist uses a technical pen or a crowquill pen, but most artists will prefer one over the other, technical pens for their consistency or crowquills for their flexibility. When either one is well used, the resulting differences are negligible.

Fine illustrations can also be prepared in pencil and reproduced as halftones. While the cost of printing is greater, the artist may work faster and therefore less expensively in pencil. Tone drawings can be aesthetically more pleasing and can provide more detail if well done and well printed. However, a line drawing may be perfectly adequate and easier to create, and is certainly more reproducible. Before any pencil work is submitted, the printer should be consulted to insure its reproducibility.

Pencil can be used on drafting film (Fig. 12). The microscopically coarse texture causes the pencil tone to come out much darker than when worked on paper. The printing process is basically the same as for photographs but good quality reproduction of pencil work is difficult because of the range from paper white background to dark gray. When a pencil drawing is printed by most reproduction methods, a gray background will surround the image (unless it is stripped out by the printer, which is an expensive process), but this does not necessarily detract from it. For best results, pencil drawings should be drawn with little or no planned reduction. The drawing must be clean and fixed with an acrylic spray before shipment.

Another method of creating effective halftones involves the use of coquille board (Fig. 15), a paper embossed with tiny bumps. A black pencil or litho crayon creates a texture when rubbed over the dots that resembles stipple work. Pen and ink can be used to outline or highlight parts of the drawing. Because of the coarseness of the texture, the artist should work with at least two times the printed size, and have the artwork photographed with high-speed litho film and submit the photograph rather than the original.

Maps

Botanical cartography, like botanical illustrations, must necessarily be educational and scientific, aiming as it does at the accurate presentation of verifiable data about natural phenomena, but it can also be artistic and give pleasure to the eye as well as material for thought.

—Stearn (1951; p. 64)

The most common use of the map in systematic botany publications is for showing distributions of taxa. Maps may also show localities of study plots, boundaries of floristic provinces, vegetation zones, plant communities, and such climatic details as distribution of rainfall or temperature averages.

Base Map

In maps, as in plant illustrations, care must be taken not to obscure the botanical information by too much detail. The map should be simple enough that it can be understood without having to study an elaborate explanatory legend. It is better to use two maps that are easily readable than one that is overloaded. The area projected should be immediately identifiable; sometimes it is useful to include an inset map identifying where on the face of the globe the detailed area is (Fig. 16).

The first step in the preparation of a map is to make a preliminary sketch with all the necessary map features, data points, and legend material. This can be done on a sheet of tracing paper placed over an atlas map. Then compile a list of the features to be shown, including the map title, the legend material, and the names of geographical features and habitations listed in order of importance. This list should be carefully thought out and have correct spelling and diacritical marks. If the map is to be drawn by another person (an artist) include any other explanatory notes with this list. The clearer the instructions the quicker and less expensive the artwork.

The next step is to establish the image area the map will occupy, and then find a suitable map from which to adapt or copy the base features. It should be at a scale allowing for about one-half reduction to printed size. Working at twice the printed size allows comfortable room for adding detail. You may have to enlarge or reduce an existing outline, topographic, or road map and redraw it, or you may be able to trim down and re-label an existing outline map, adding new borders. The most effective map is one tailored to the particular needs and rarely does a standard printed outline map work as effectively as one specifically designed for its needs (Fig. 16). Some projects may require standard base maps for all data, such as *Flora Neotropica* (Gentry, 1980). If no

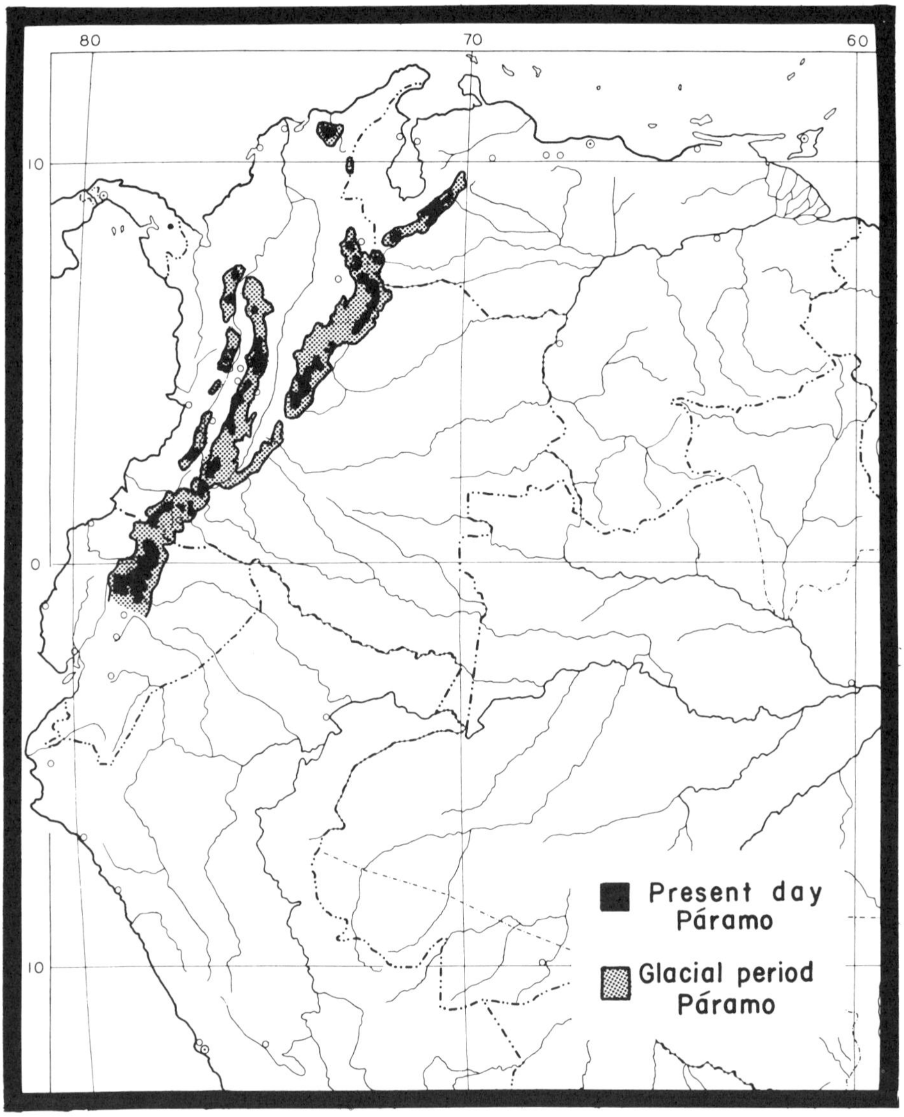

Fig. 16. Examples showing selection of base map. The right map is tailored to fit the particular need with an inset map for orientation. The left map is drawn on a store-bought outline map, which does not suit the particular need as well.

suitable base map exists, one can be drawn, tailored to accept a wide range of information, and reproduced by an offset printer. The reproduction must be on good quality paper if it is to be further elaborated on and submitted to the printer.

Drawing the Map

Once decisions have been made on how much detail to include, line patterns and weights (thickness), the style and size of lettering, and the use of patterns must be worked out.

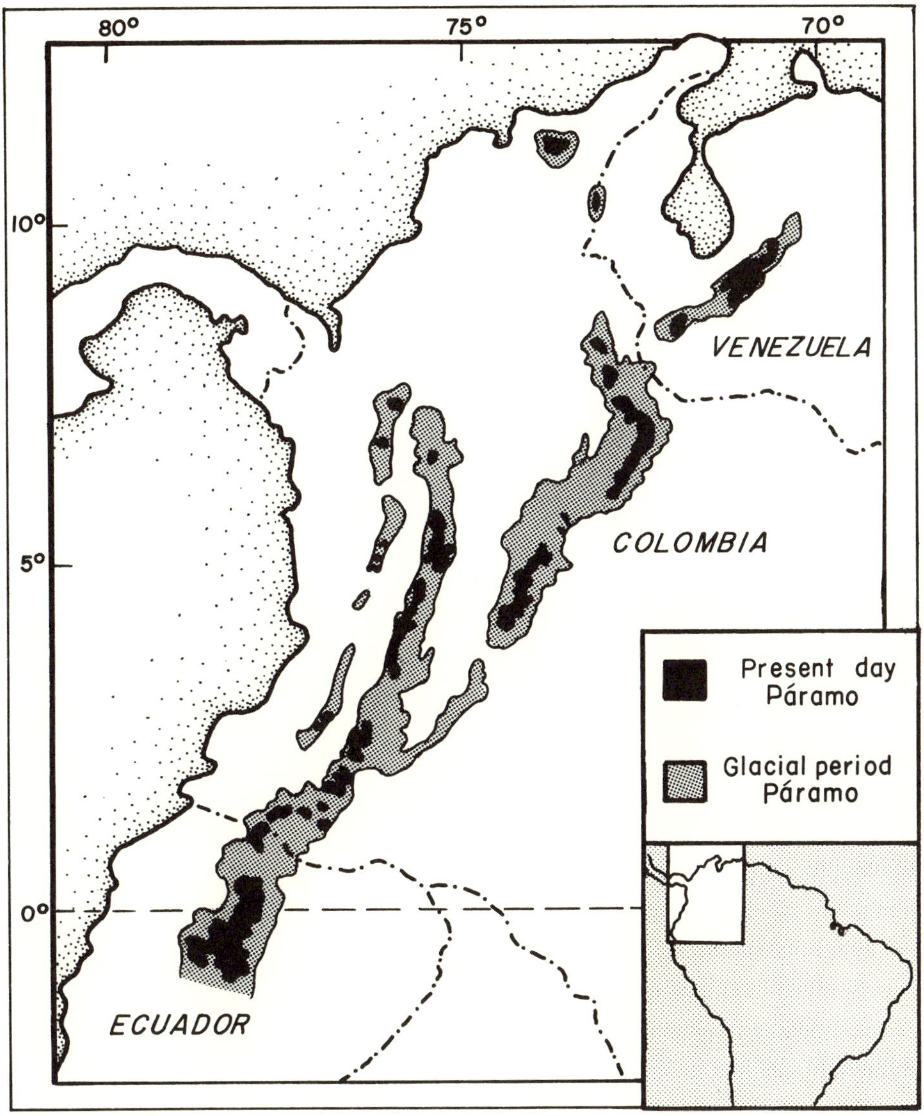

Line Features.—In determining line weights you need first to establish a hierarchy, so that the features of greatest importance dominate the map. Too heavy a line for background details may mask the relevant data. Lines indicating coastlines, rivers, bodies of water, and political boundaries should be differentiated by their widths or patterns of interruption. Natural features (except intermittent streams or dry lakes) are usually shown by continuous lines, with coastal outlines the heaviest. Political boundary lines can be differentiated by dash or dash-dot patterns.

In some instances borders of geological or topographical features can make plant distributions more meaningful. Some examples are: the inland margin of the coastal plain, the eastern border of the Sierra Nevada uplift, and the southern limit of Pleistocene glaciation (Fig. 17). A different line pattern or tone screen can distinguish these from political boundaries and water courses.

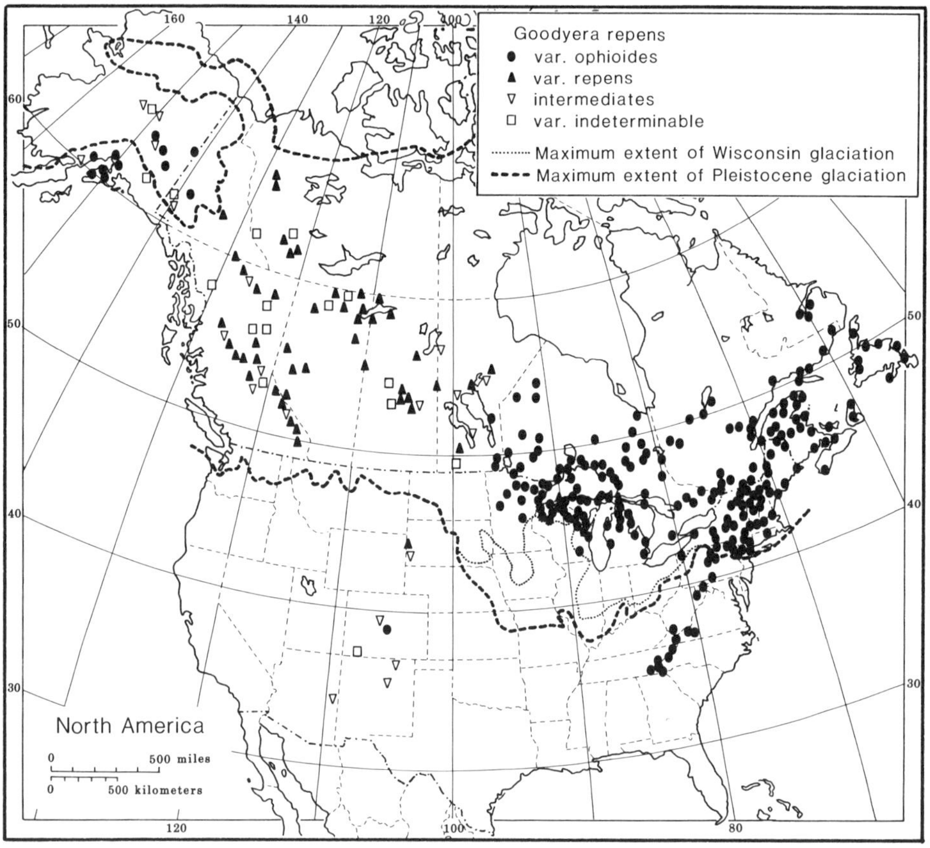

Fig. 17. Example of a map showing limits of Pleistocene glaciation in North America and the distribution of *Goodyera repens.* Adapted from Kallunki (1976, p. 59), and used with the permission of the author.

A map can express three dimensions with elevational contour lines. This enables the author to demonstrate distributional gaps correlated with altitude. One should be careful to choose intervals that will not become too dense when reduced.

Lettering Considerations.—Type sizes should be carefully chosen in accordance with the features to be emphasized on the map, the amount of open space, and the relative importance of place names. As always, anticipated reduction must be considered.

It may be helpful to lay out all of the lettering before drawing the map features so you can see what details have to be omitted to make it readable. Various options for placement of lettering may exist, such as city names to the left or right of its locale. When lettering will be confused by map detail, it may be best to mask out part of that detail to avoid overlap, or connect the lettering to its locale with an arrow. Typewriter correcting tape is useful for masking out space. The Kroy lettering machine can be used with either an opaque white tape or a transparent tape.

There are few specific rules governing lettering in cartography, but there

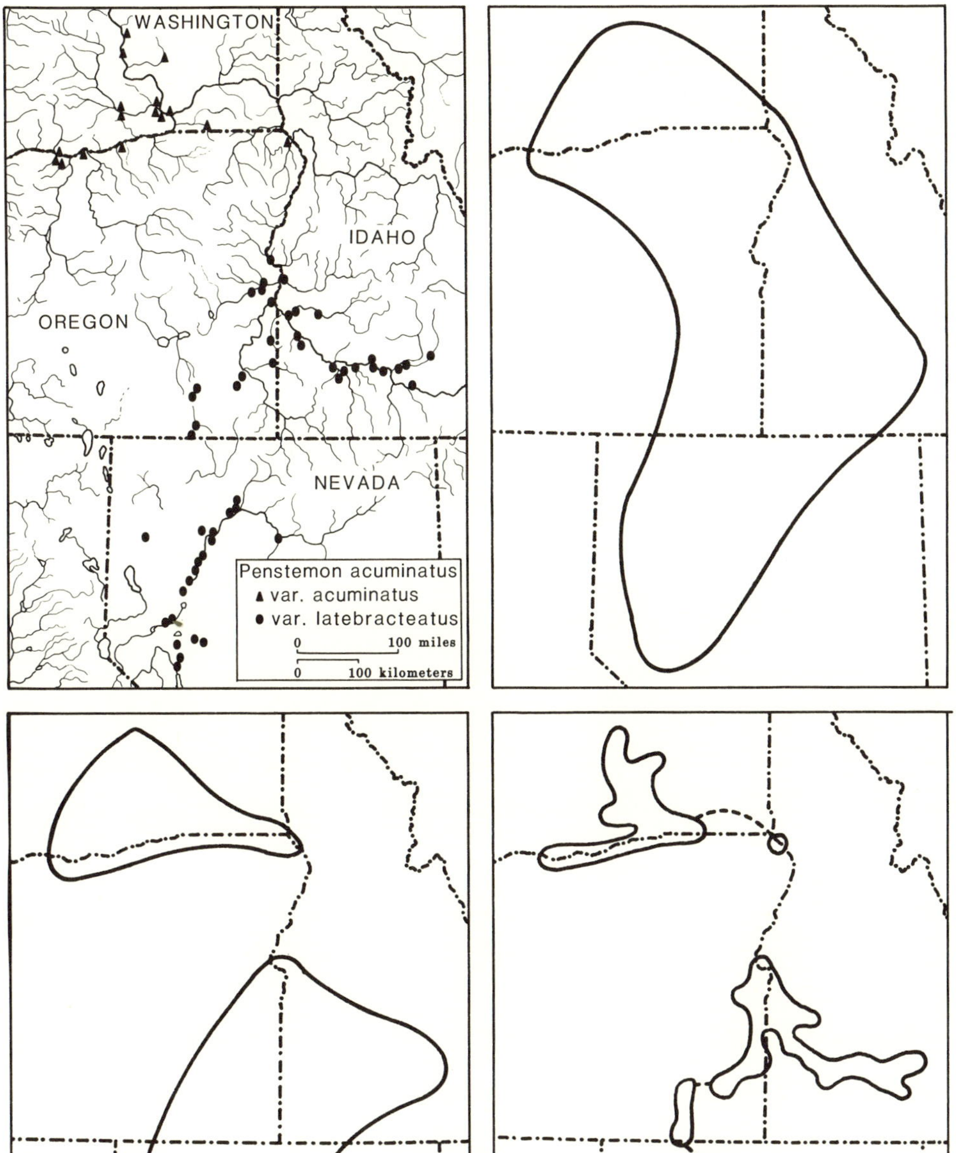

Fig. 18. Maps of the interior of western United States, each showing the distribution of the two varieties of *Penstemon acuminatus*. The upper left is a dot map showing locations from which herbarium specimens were collected. The remaining three, using the border method, show different degrees of accuracy depending on intended use or mapper's knowledge of taxonomy and topography. Adapted from Holmgren (1979, p. 235).

are some common practices. Type size can be varied to indicate the magnitude of political features. Capital (cap) letters can be used for important names, cap-small cap for intermediate names, and cap-lower case for the least important ones. Names of large areas such as countries can be spread out across the space, usually using capitals. Italic lettering is usually used to distinguish water features (seas, lakes, rivers, etc.), possibly because the slanting or curved letters suggest fluidity.

The function and aesthetic harmony of maps will be enhanced by adherence to certain standards, although these are not hard and fast.

1. A label should be placed either entirely on land or on water.
2. Where lines or tones conflict with names the line or tone should be interrupted rather than the name.
3. Lettering should not be upside down.
4. Diagonal lettering should not be set in straight lines, but rather in a slight curve, to avoid the appearance of a printed label sloppily pasted on the map.
5. Curvature of words should be graceful, with no sudden change in direction. When labeling a river, choose a stretch that is fairly straight.
6. Lettering that refers to point localities should, when possible, be placed to the right of the point or slightly offset to the right and above or below the point.
7. A name of a line feature (e.g., river, boundary, road) should be placed alongside and parallel to it.
8. If vertical lettering is necessary it should read upward on the left side of the map and downward on the right side.

Words lettered on a curve, such as the name of a river, require special attention. A curved guideline should be drawn in blue pencil. With press-on letters align each letter perpendicular to the curve. With a lettering set (such as a Leroy) mark the locations for each letter and align the scriber (centered on the letter to be traced) at a position perpendicular to the curve, changing the angle for each letter. With the Kroy lettering tape, if the curve is subtle enough, a deep nick can be made in each letter-space and the tape bent to fit the curve, or individual letters can be cut out and individually placed.

Tone Screen.—A tone screen pattern can be used to distinguish regions on maps where it would not obscure patterned botanical information. A pattern can be used to separate water from land, but it should be subtle enough to not draw attention to the water. When zones or regional areas of a map are distinguished by different tone patterns, adjacent values should contrast with at least a 30 percent difference (see section on tone screens in tools and materials).

Legend Box.—The addition of a legend box enhances the visual impact and usefulness of a map. A map title, key to the symbols or tones used, and a scale can be included (Figs. 17–19). The legend box within the map is preferable to a description in the figure caption, which is once removed from the map. If the data-point symbols have to be explained in the figure legend, use standard symbols that the printer can set: ○, △, □, ●, ▲, ■, or that can be easily described (open circles, closed squares, etc.). As with plant illustrations, a scale bar on the map is preferable to a scale ratio in the figure caption because its relationship to the map is not affected by reduction.

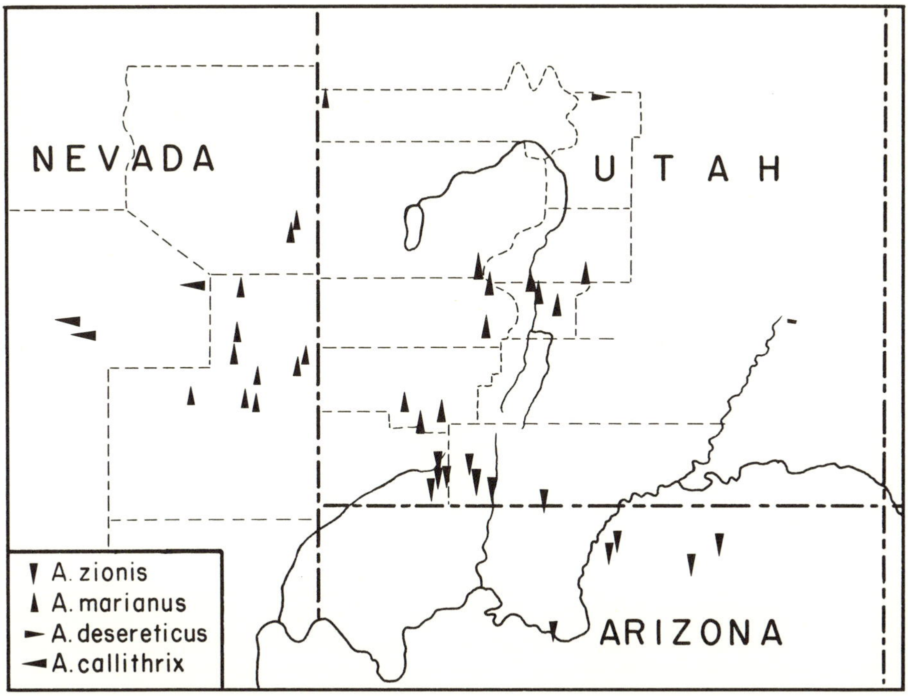

Fig. 19. Dot map of western United States showing the distribution of four species of *Astragalus* using isosceles triangles in different orientations to distinguish the species. Modified from Barneby (1964, p. 632), and used with the permission of the author.

Other Refinements.—If the north direction is not towards the upper edge of the map an arrow indicating north should be included.

A border usually improves a map. Latitude and longitude coordinate numbers can be placed outside a border or within a double border (Fig. 17), and if the lines are not drawn across the map, their positions can be indicated by small ticks inside the border line. Press-on line tapes, such as Letraline Graphic Art Tapes, are useful for borders. If a base map is trimmed down and used directly, the missing borders can be added with a matching tape size. If several maps are shown on one plate they can be separated with tape lines.

Materials

Drafting film is especially well suited for maps because the work can be done directly on top of the sketch or base map. Alterations and corrections of errors are easy to make with a plastic eraser or scalpel, so that boundaries can be changed or names eliminated if the map becomes too crowded.

Distribution Maps

The distribution of taxa is an important part of descriptive taxonomy and plotting it on a map provides an excellent visual aid for the reader. Distribu-

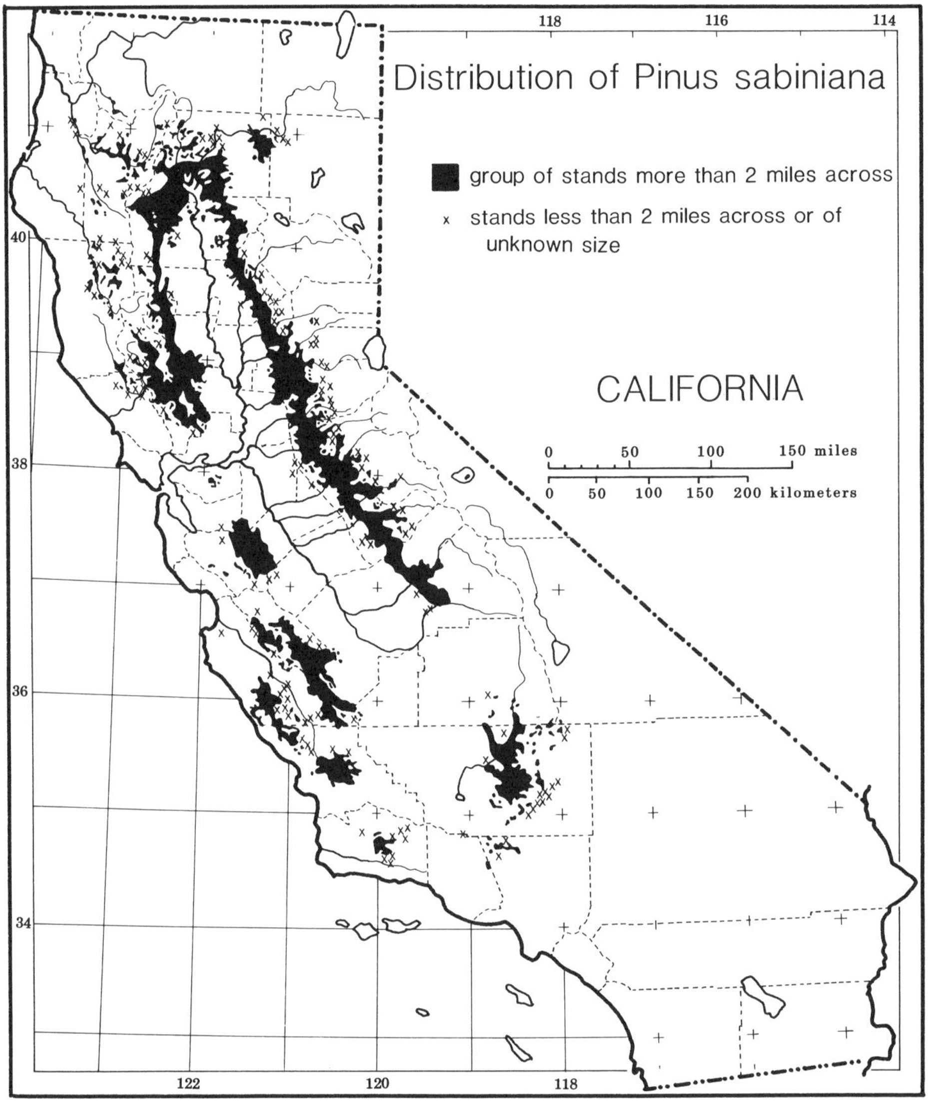

Fig. 20. Map of California showing the use of shading and dots to distinguish extensive communities from small outlying populations of *Pinus sabiniana*. Map modified from Griffin and Critchfield (1972, p. 103) with the consent of the senior author.

tions can be represented in two basic ways: the boundary line method and the dot method. The boundary line can be further varied by blocking it in. Combinations of boundary lines and dots, or blocking in and dots can be used to advantage.

Boundary Line Method.—The outlining of the area of distribution by a connected line is referred to as the boundary line method. It is not precise because the peripheral distribution of a taxon is often sparse. The boundary line method cannot indicate whether a taxon is rare or common, but it can display its entire range. A knowledge of the habitat, the topography, and morphological

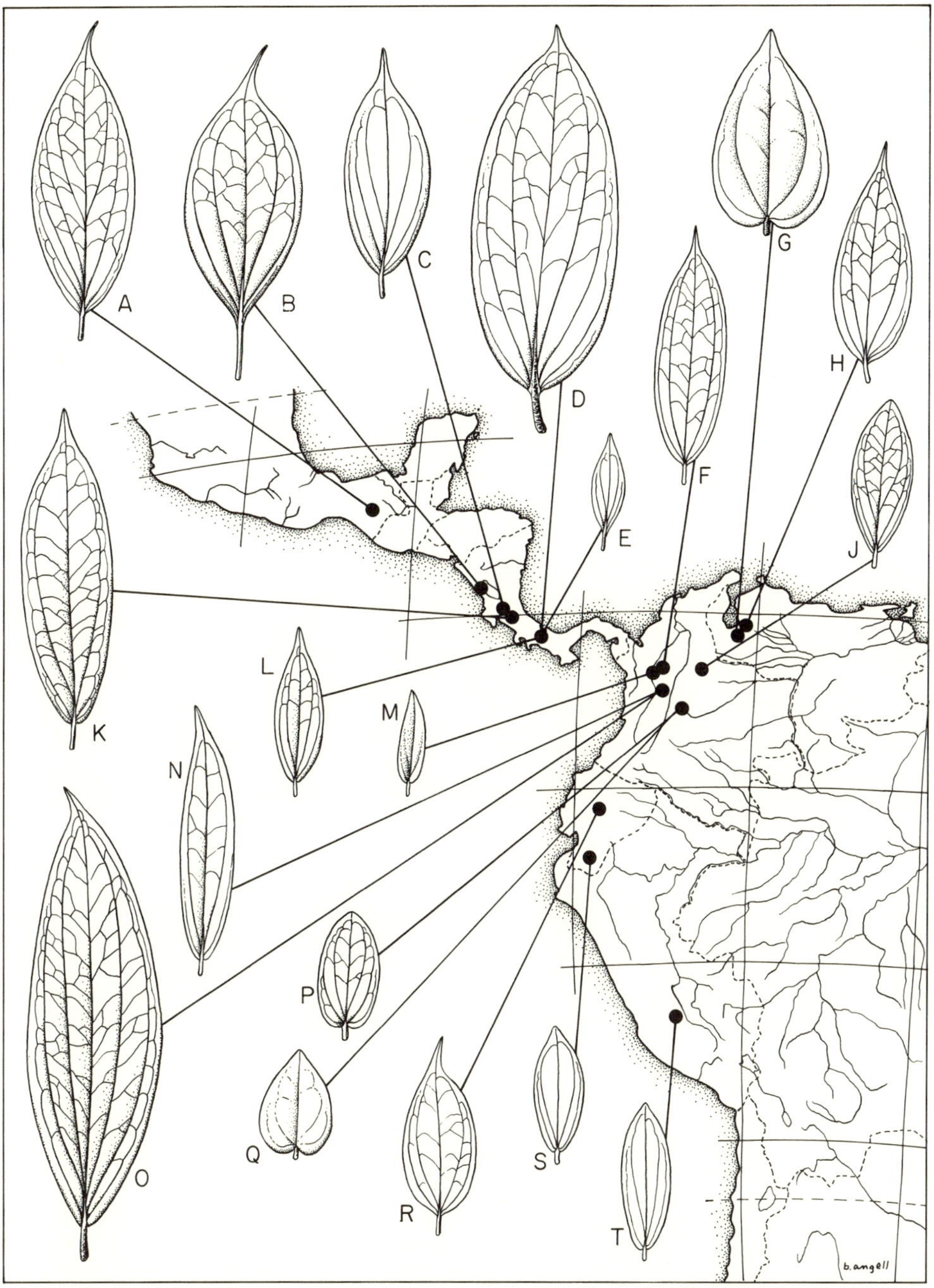

Fig. 21. Map of Tropical America using a combination of dots with illustrations showing geographical occurrence of leaf forms in *Cavendishia bracteata*. From Luteyn (1983, p. 151) and used with the permission of the author.

variability of the species helps in determining the most accurate placement of the line (Fig. 18).

This method is useful in summarizing the distributions of two or more related taxa on a single map. Each species area can be identified by a number corresponding to that used in the text. To make each species more distinguishable, the boundary lines can be drawn with different kinds of lines, such as solid lines versus lines that are broken, dotted, zigzag, dash-dot, or dash-dot-dot, etc. To further enhance the boundary line map, the enclosed area can be shaded with a tone pattern. Different patterns of shading film can distinguish between taxa. If overlapping taxa are shown, line patterns should be selected that can be overlapped and produce a uniform third pattern.

Dot Method.—The dot method is often the best way to show distributions. Dots placed in precise localities documented by herbarium specimens indicate what the author knows and suggests what may remain to be learned. It shows only the supporting data and makes no attempt to fill in the gaps.

Variations on symbols are used to distinguish different taxa, or to distinguish certain collections, such as the type specimen or plants that have been examined cytologically. Circles, triangles, squares, diamonds, and stars can be differentiated as solid or open shapes or by stylizing them with projecting arrows or appendages. A useful way of varying an isosceles triangle or an arrowhead is by orienting its point in different directions (up, down, right, left) (Fig. 19).

A method sometimes used for placing dots is the grid-square method. The presence of a species is indicated within a specific grid by a solid dot. This method is used in maps for *Flora Neotropica* where each grid is one degree square and in the dotting of the small, evenly spaced counties in the *Atlas of the Flora of the Great Plains* (Barkley et al., 1977).

In maps showing the range of more than one species with overlapping distributions, a combination of dots and boundary lines may be easier to read. It combines the accuracy of the dot method with the ease of observation of the boundary line method. For some species shading may indicate where the taxa occur abundantly and dots indicate less frequent locales (Fig. 20).

Pictorialized dot maps can convey information on morphological variation as related to geography (Fig. 21). The data should be simple and easily identifiable.

Graphs

I ask you why it is that some scientists will demand the newest ultracentrifuge, even if it costs $40,000, and then refuse to pay $20 to a commercial artist to draw a proper graph of the results obtained with the ultracentrifuge.

—Robert A. Day (1983; p. 157)

A graph is a pictorialized table. It must be accurate and it can also be attractive. It is often the best way to present statistical information where trends and comparisons can be readily visualized. To most effectively do this the graph should be simple, displaying the trend with a minimum of data. If actual tabulation of data is important, it should be given in a printed table instead.

Unlike botanical illustrations and maps, graphs can be prepared with little skill, often without having to use a pen. Adhesive tapes, symbols, and letters make graphs easy to produce. The graph can be prepared on frosted drafting film over graph paper, eliminating the need for a light box, or on paper with a blue grid.

Designing an effective presentation does take skill, however. A graph should be carefully thought out. There are different ways of graphically presenting information, through data points, line curves, or bars. The data can be plotted vertically or horizontally, whichever takes less space and is easiest to read. For more information on artistic execution and choice of method for the kind of data see MacGregor (1981).

Line Graphs

Curve lines, whether they connect data points or display a trend, should be drawn evenly and with uniform thickness. If you are not confident with a pen, press-on line tape can be used. Letraline has flexible tape in several widths that can be easily bent into curves. Symbols for data points should be the standard circles, squares, triangles, diamonds, or ×'s. If more than one set of data is to be illustrated, curve lines can be varied as dotted, dashed, or solid, or the data points themselves can be varied (Fig. 22).

When solid lines cross, one of them should be interrupted. The axis should extend to only one increment beyond the final data point. The labels and calibration numbers of each axis should be placed outside the lines. A numbered scale should start at zero. If this is not suitable it should be clearly indicated. The interval marks (small ticks) of the axes should not cross the line; they may extend either into or out of the data area. The interval marks should merely serve as reference points. They should not be too numerous and not every one need be numbered.

A key to the symbols or curves should be included in a legend box within

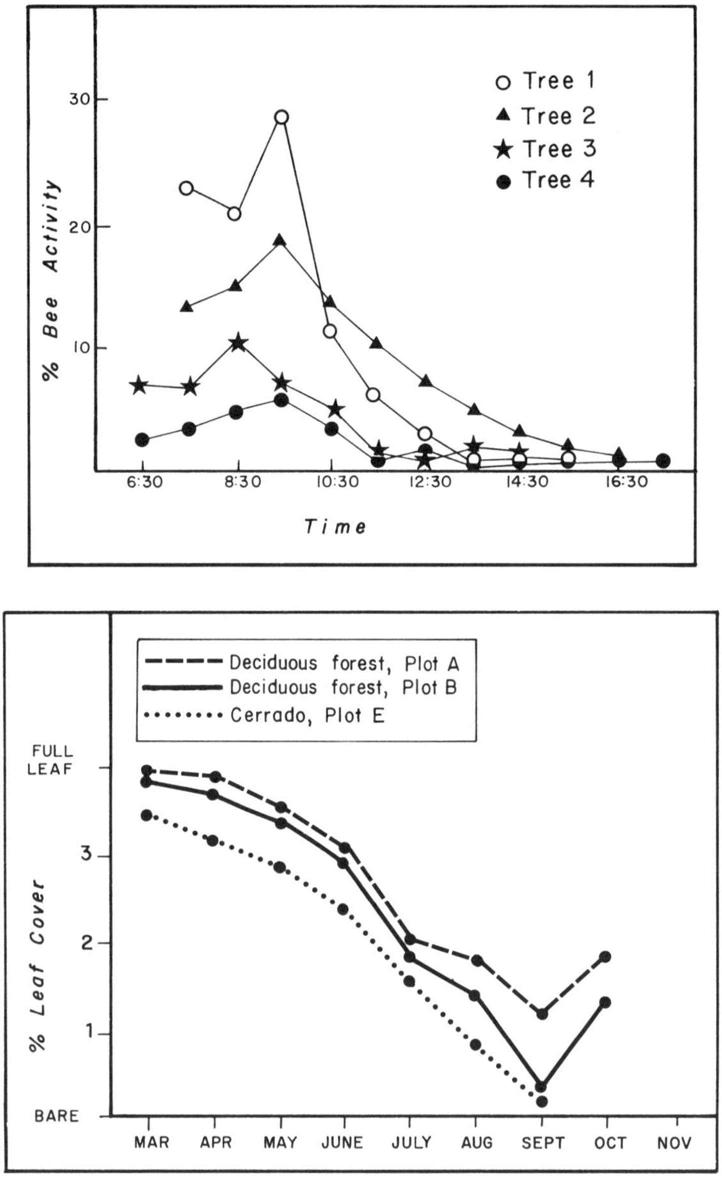

Fig. 22. Two graphs showing variations on data points and lines. The upper graph is adapted from Mori and Pipoly (1984, p. 339), and the lower one is adapted from Prance and Schaller (1982, p. 245). Used with the permission of the senior authors.

the axes if there is space where no data appear (Figs. 22 and 23). A key to the symbols within the illustration makes it easier to interpret than if it has to be given in a printed figure caption below.

The thickness of lines used in conveying information should be selected according to their importance. The data points and their connecting lines

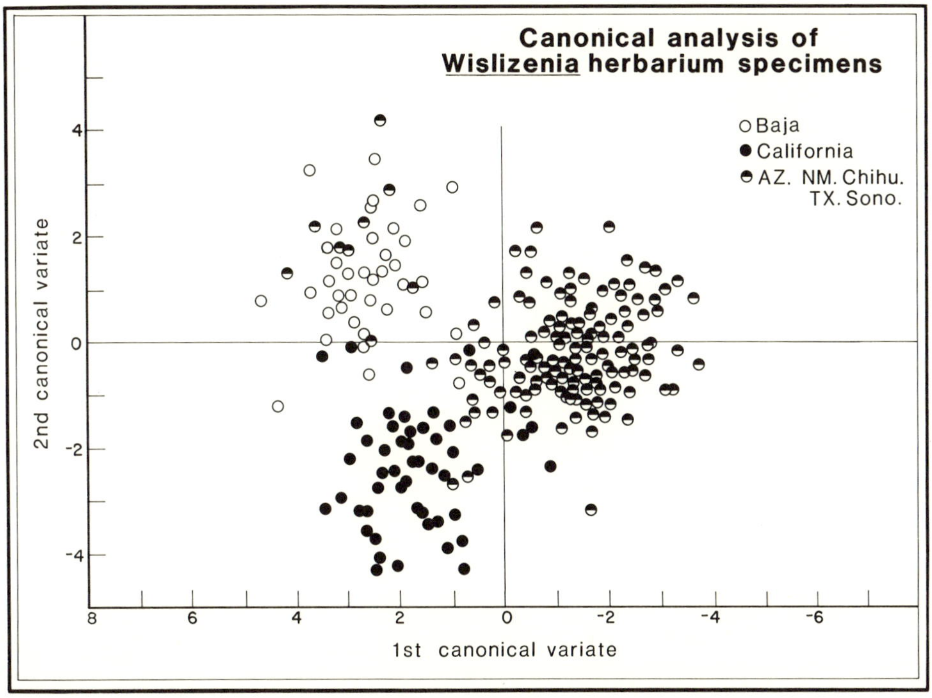

Fig. 23. Graph showing canonical analysis using open, closed, and half-closed circles to distinguish geographical entities. Modified from Keller (1979, p. 340) and used with the permission of the author.

should be the boldest, the axis labels and lines less bold, curve labels less again, and finally the interval marks and numbers the least bold.

In labeling the graph the largest letters are used for the title, the axis labels are smaller, and the figures on the axes and the letters in the key to symbols are the smallest.

Bar Graphs

As with line graphs, bar graphs should be simple and easy to read with an emphasis on the trend rather than the specific data (Fig. 24).

Tone patterns within bars can enhance graphs by making the information stand out and allow for emphasis within the graph. Dot patterns are preferable to line patterns for shading bars; line patterns can cause optical illusions. Adjacent values should contrast well, with at least a 30 percent difference (see notes on tone screens in section on tools and materials). Make either the most important or the smallest bar darkest. If the bars are spaced apart the space between should be narrower than the bar. If one bar is disproportionately longer than the others it can be broken and indicated as such. The corresponding axis should be broken also (Fig. 24). An alternative to labeling interval lines is to label the bars with actual amounts.

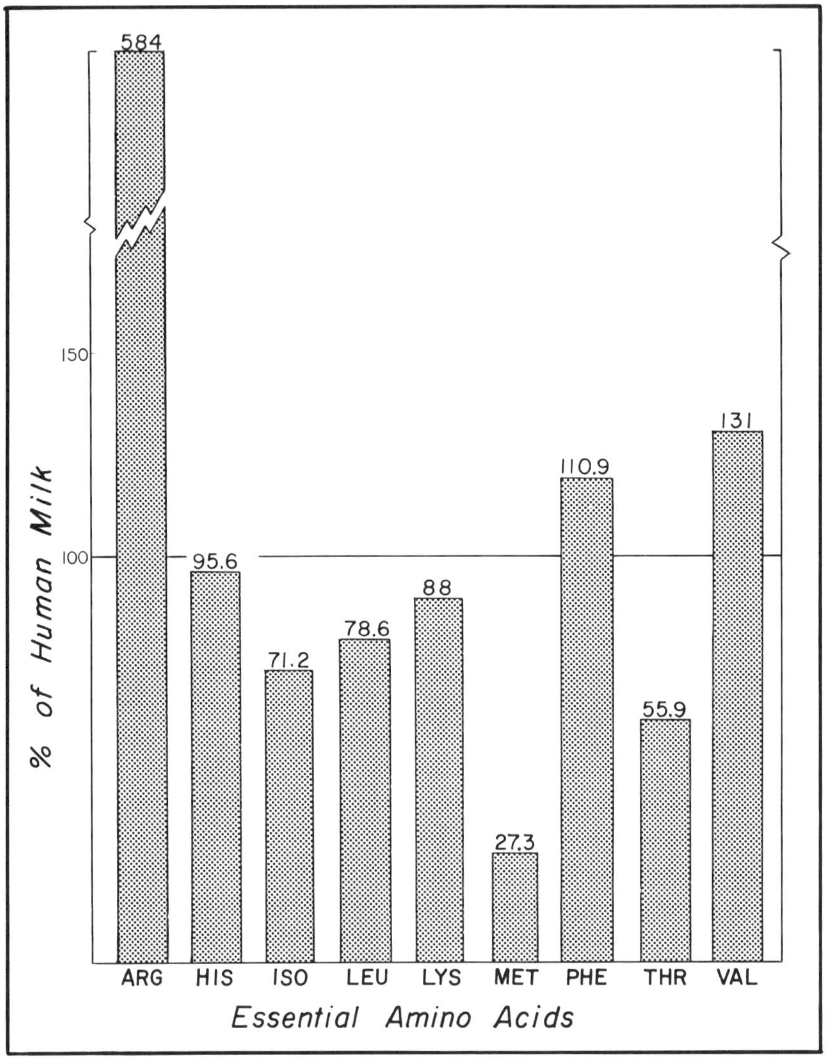

Fig. 24. Example of bar graph showing an acceptable method of breaking disproportionately longer bar and showing method of labeling actual amounts. From May et al. (1985, p. 124) and used with the permission of the authors.

Photographs and Other Halftone Work

Halftone (continuous-tone) artwork, such as a photograph, painting, or pencil drawing, is reproduced as a halftone. This takes more steps than line art in reproduction and the cost is therefore greater. The cost of a halftone can be justified if it shows more than that which can be illustrated with a line drawing.

Illustrations of vegetation types, habitats, scenery, and the living plant are best served by photographs. To show three-dimensional depth the photograph should have a good range of tones from very dark to very light, with details both in shadows and in highlights. The subject should be easily recognizable and in precise focus, with sharp outlines and clear details. The background should be simple with no distractions and contrast with the subject. An indication of scale should be included.

A photograph of a herbarium specimen seldom illustrates more than vegetative form and branching patterns, if that much. Details of the plant are often lost in obscuring vestiture, or overlapping parts, or are invisible due to poor contrast. Diagnostic features are usually better shown in a line drawing. A photograph of a herbarium specimen can be useful for publication if it portrays a specimen of historical or nomenclatural importance. When photographing a herbarium specimen, one way to show all the features is to minimize the shadows by using diffused light. The most important information may well be the collection label and annotation labels on the sheet. Be sure to include a scale (always metric in scientific works).

An electron micrograph should be shot at about the size it is to be reproduced so that there is no need for reduction. Before exposing the film be sure that the subject will be the right size, allowing for the cropping of any extraneous parts.

Photographs of cellular sections, SEMs, habitats, etc. can often be grouped together into a composite plate. It is best if the photos within a plate all have similar background lighting, camera angle, and scale to create a coherent, easily understandable image (Fig. 25).

For publication it is best to submit glossy prints. Matte and textured finishes are less desirable because they are difficult to reproduce.

Enlarged black and white prints from 35-mm color slides are not always satisfactory. The botanist planning to use photographs in publications may want to carry two cameras, one for black and white prints for publication and one for color slides for other uses.

The inexperienced photographer may want to go to a professional who has good equipment and a good knowledge of how to enhance the subject through selection of films, lenses, filters, aperture settings, shutter speeds, and focal lengths. A good reference text for doing it yourself is Blaker's *Handbook for Scientific Photography* (1977).

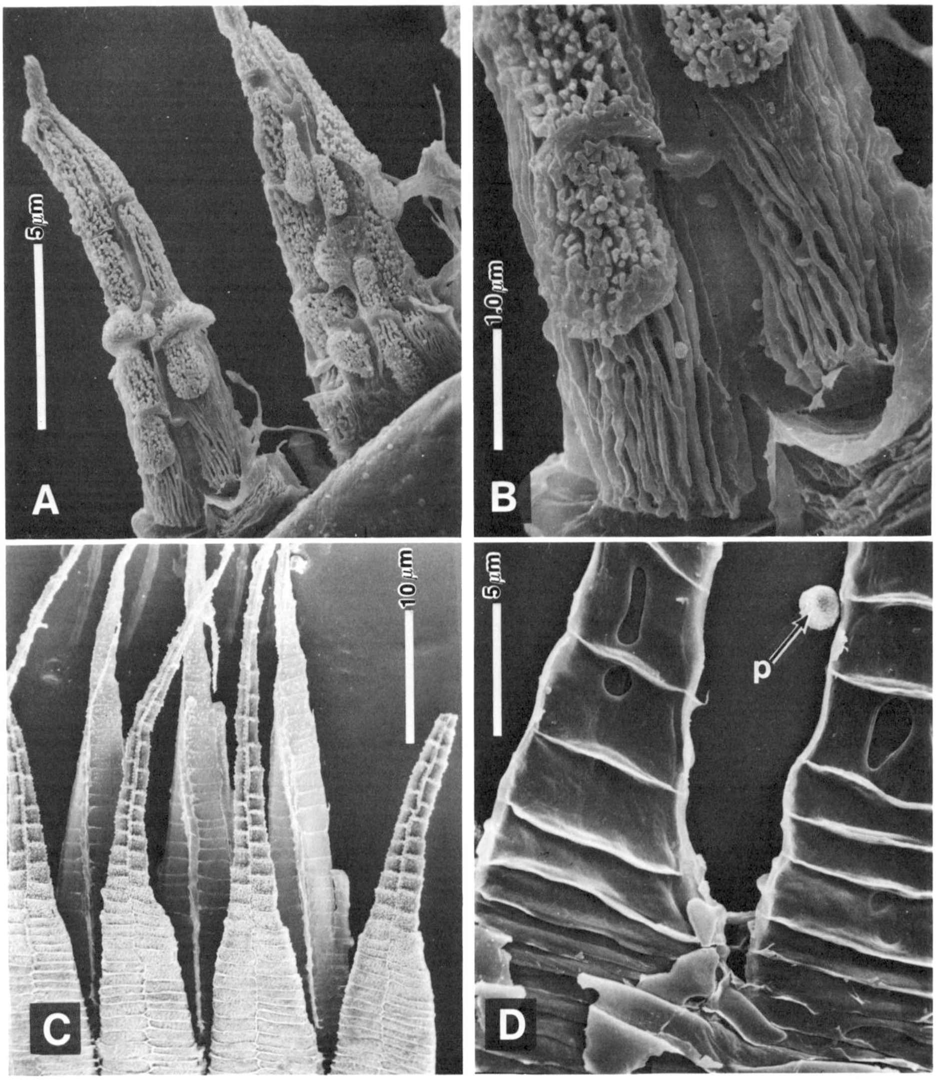

Fig. 25. Four photographs cut and mounted snugly together allowing the printer to scribe a neat white hairline between. The scanning electron micrographs of moss peristomes were supplied by William R. Buck.

Halftone reproduction involves photographing the artwork through a "half-tone" glass screen to create a series of black dots on white paper, which optically appear as shades of gray. Areas of the picture that appear light gray have small black dots and darker areas consist of larger dots that may touch or overlap each other.

The halftone screen consists of two plates of glass cemented together, each with a pattern of equally spaced engraved opaque lines on the inside surface, crossed at right angles to give a series of evenly spaced openings. The screen is placed between the negative and the film being exposed. The more intense the light the larger is the area exposed through the opening by the halo effect and consequently the larger the dot. The screens differ in mesh size from 65 to 300 openings per inch. The greater the number of dots, the more accurately the reproduction resembles the original continuous tone.

If the halftone art is of two-dimensional fine structure with important but poorly contrasting details, be sure the publishing journal uses halftone screens with at least 155 lines to the inch and smooth-finish or coated paper. A 300-line screen produces photographs with no visible dot patterns to break up the image, so subtle and delicate details can be reproduced. Not all printers can produce quality 300-line screen, and not all journals can afford the cost. The fine screen can only be used for printing on a smooth coated paper surface. A coarse screen is better for rough finish such as newsprint or rough cover-stock.

Previously printed halftones taken from published sources already have the screened dot pattern of the halftone on them and when reprocessed the new picture is of poor quality and extra cost is encountered. If the screen used originally was coarse and the printed dot pattern is of solid black against a clean white background (as seen under magnification) it can be reproduced as line art without rescreening.

The use of halftone art with line art in the same plate may be effective, but if properly prepared by the printer the cost can exceed that of separate line and halftone preparations. Some printers will shoot such combination plates as halftone art, but the white background in the line copy will take on a light gray appearance and the lines will not have a crisp even outline. If the line art is just the identifying numbers or letters it is satisfactorily reproduced if it is kept within the bounds of the photograph, in which case it is shot completely as a halftone.

Pencil drawings and wash paintings are reproduced as halftone art and the white background in the original will usually reproduce as gray unless the printer carefully masks it out. However, as the technology of printing processes improves, better reproduction of pencil will become possible.

To insure good quality reproduction of halftones follow these guidelines.

1. Submit photographs at the same size they are to be reproduced.
2. Keep fingers off the surface of the photograph; oily fingerprints can damage the film emulsion and will likely show up in the print.
3. Always cover the photograph with a protective overlay of drafting film or vellum. Do not use drafting film if pressure-sensitive labels have been used.
4. Never use paper clips on photos. They could dent the paper or emulsion.
5. Photographs should be mounted on a board that is large enough to allow space for editor's notes to the printer.
6. Technical pens and magic markers can be used directly on the back of an unmounted photograph, but the best way to label a photograph is to type information on a self-adhesive label that can be affixed to the back. Never use a ball-point pen for writing on an unmounted photograph.
7. If there is something in the photo that requires special attention by the printer, circle the area on the transparent protective overlay with a soft-tipped marker, being careful not to press too hard. Another method is to write the instructions on a photocopy and submit the copy with the original photo.
8. Crop marks should be indicated on the protective overlay or in the white margins. The printer can get cleaner margins from a photo which can be

cropped by him than one trimmed by the author to exact size. Sometimes a cutting tool will leave ragged edges that may flake.

9. When mounting more than one photo to a plate, cut the edges as squarely as possible and butt them snugly together leaving no white space between. The printer can scribe a neat white hairline between the photos (Fig. 25). He will do this in black to the negative of the screened photograph, which results in a crisp, clean, white line.

Good color reproduction is complicated and expensive. Color printing involves four printing operations that must be carefully coordinated. All normal colors are reproduced by varying combinations of blue, red, and yellow process inks, and black is used to provide depth and additional detail. The high cost comes from all the processing stages, which include photo-separating the artwork through screens to eliminate all but one color four different times, and printing the artwork four times with each color perfectly registered (aligned). The printed artwork will have an overlapping pattern of colored dots visible with a hand lens.

As technology advances, the cost of color reproduction decreases and may soon be within the price range of some journals. Printing processes vary in their treatment of color work with some depending on flat artwork and others on flexible art or transparencies that can be arched for laser scanning. The artist should check with the printer before beginning his or her work.

Preparation for Shipping

Line Art

Some editors prefer reproductions of artwork for publication, or may not want original illustrations until the paper has been accepted for publication. In place of originals a quality photograph or photostat (not photocopy) can be submitted. Artwork done on drafting film or vellum, or involving a lot of paste-up work should be submitted as a photostat to eliminate risk of damage.

A photostat (or stat) is an inexpensive photographic print that reproduces line work quite faithfully. It is a black and white opaque copy made from the original by a stat camera. The process uses a high contrast exposure paper that holds only black on white. The stat camera can reduce or enlarge the image to any reasonable size. A commercial graphic studio, blueprint firm, or photography studio can usually make photostats. The print should be made on matte finish paper. If you wish the stat to last a long time you need to tell this to the printer so he will rinse it thoroughly after running it through the developing solutions. A PMT (photomechanical transfer print) is similar to a photostat but the resolution is not quite as fine.

Mounting Plates

When the original drawing is required, preparation for shipment should be carefully done to insure against damage. Original artwork should be mounted on a fairly stiff board before shipping. The artwork should be bonded or taped to the mounting board with at least one inch margin at either the top or bottom allowing for the editor's identification and instruction notes. If the illustration might be used for exhibit or archival storage the mounting should be a temporary one.

Sometimes artwork is used right away to illustrate a new species and years later incorporated into a monograph or general flora. While it is in storage, even for a couple of years, tape, glue, and poor quality mounting board can do their damage. The artwork should be separated from its harmful materials as soon as it is returned from the printer.

Photographs can usually be replaced so mounting does not present as much of a problem. Photographs can be mounted on a stiff board such as Bainbridge board or 16-ply poster board with rubber cement, wax, or by dry mounting. A set of copies can be retained for archival storage.

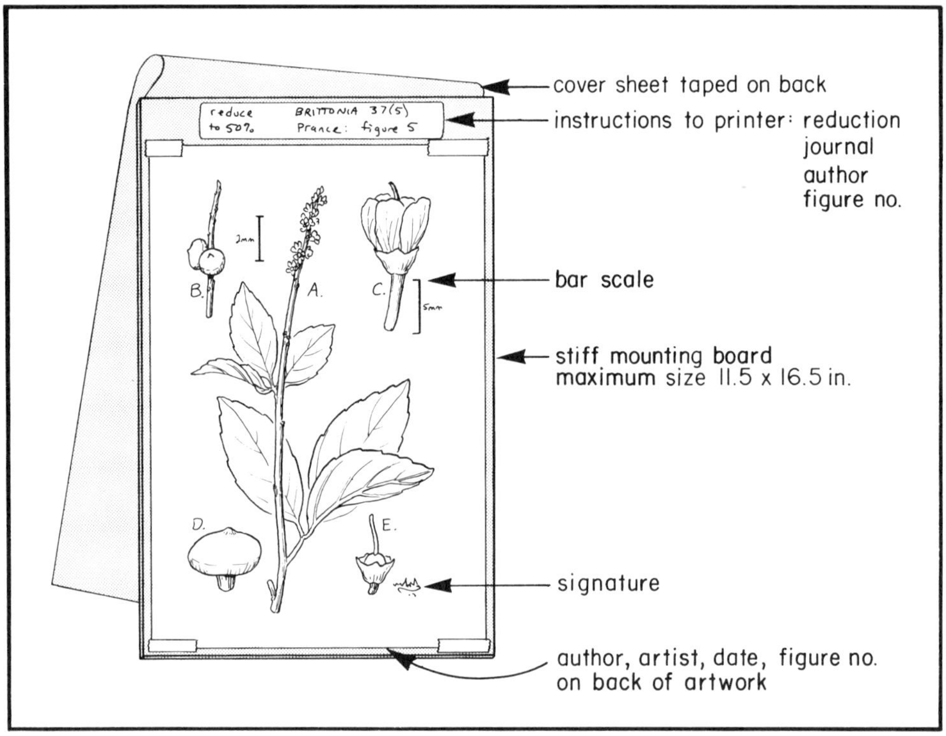

Fig. 26. A properly mounted and labeled illustration.

Protective Cover

All art should be protected with an overlay sheet (Fig. 26). A protective overlay of vellum (100% rag tracing paper) is recommended. Drafting film is good also, as long as no pressure-sensitive letters were used. If the overlay is transparent it need not be opened except for close examination and for photographing by the printer.

If the upper margin of the mounting board is sufficient, the overlay sheet can be taped onto it. A preferable way is to fold an inch or two of the overlay over the back of the board where it is taped or rubber cemented. Secure it along one edge only, so it can be examined with ease.

Identification

Identification of artwork should serve three purposes.

1. The editor and reviewers must be able to identify the illustration. This should include the author's name, manuscript figure number, and journal name and can be put on the back of the board.
2. The printer must be able to identify the plate. If the information is on the top of the plate, neatly typed or written, the printer can photograph it on the same image as the plate. The editor may have his or her own

system of labels and can transfer the author's information directly onto them.

3. The artist should be identified on the back of the original so that it can be catalogued in archival collections with species name, artist's name, and date of completion. The cover sheet is often torn or removed so it should not be used for identification.

Wrapping

Oversize illustrations can be easily damaged in shipment. A maximum size of 11½ by 16½ inches (30 × 42 cm) is recommended because most authors and editors of plant systematic journals will have herbarium sheet-sized cardboards and shipping boxes available. If you have difficulty wrapping your illustration for shipment to the editor, there will likely be difficulties rewrapping it for shipment to the printer and for return shipment to you.

Illustrations should be placed between two pieces of corrugated or heavy cardboard sheets cut slightly larger than the plates to protect the corners. If there is more than one plate, stack them face up and cover the stack top and bottom with heavy cardboard. Make sure the plates are held tightly in place so they will not shift around.

An outer covering of heavy brown wrapping paper or a sturdy manila envelope should be used. Camera stores sell photograph mailing envelopes, which are made of durable paper with pre-cut cardboard. The U.S. Post Office has good padded envelopes, which give added protection on top of cardboard, but the padding does not effectively replace cardboard. Artwork should not be shipped rolled. Tubes are too easily damaged, and the artwork is difficult to deal with once it has assumed a rolled state.

Be sure your address and the mailing address are attached inside just in case the outer labels are rendered illegible.

Shipping

As protection against damage or loss of artwork in transit, keep a good photostat or photocopy to use as a substitution or as an aid in redrawing it.

Art sent through the U.S. Mail should be registered. It presumably will be handled more carefully and can be traced if lost. Artwork mailed by Parcel Post can be registered, insured, and given a Return Receipt (which assures you of the date the material was received and by whom). Express Mail and similar express delivery services do not cost much more and the overnight delivery saves prolonged worry. Insurance should be for replacement value. UPS is fast and reliable, but they will not insure original artwork.

Annotated Bibliography

For those beginning to draw biological specimens we recommend the following self-teach books: Downey and Kelly's *Biological Illustration* (1982), Wood's *Scientific Illustration* (1979), and Zweifel's *Handbook of Biological Illustration* (1961). These books are broad in their coverage of scientific illustration in general. A book that beginners as well as accomplished artists should read is West's *How to Draw Plants* (1983). It specifically addresses the techniques of botanical illustration.

We only touch upon the techniques of the printer. More detail can be found in Allen's *Steps Toward Better Scientific Illustrations* (1977), Craig's *Production for the Graphic Designer* (1974), and Stone and Eckstein's *Preparing Art for Printing* (1983). These books provide important background knowledge for the preparation of art for publication, describing platemaking, typography, and printing procedures. A knowledge of these processes is helpful in knowing what problems the printer deals with when reproducing artwork.

For more information on map making and kinds of maps see: Robinson, Sale, and Morrison's *Elements of Cartography* (1978), Stearn's "Mapping the distribution of species" (1951), Davis and Heywood's *Principles of Angiosperm Taxonomy* (1963, pp. 308–321), and Brouwer's "The cartographer's role and requirements" (1983).

The preparation of tables is not dealt with here, even though they present the same kinds of problems of layout and ingenuity involved in art composition. The printer usually takes care of their preparation based on instructions from the editor. Some journals or symposium proceedings require submission of tables in camera-ready form. Tables are adequately dealt with in the *CBE Style Manual* (1983) and in Day's *How to Write and Publish a Scientific Paper* (1983).

The following is an annotated list of books with which we are familiar that might be useful to botanists and beginning botanical artists. Most of the books are available directly from the publisher.

Allen, Arly. 1977. *Steps Toward Better Scientific Illustrations.* 2nd ed. Allen Press, Inc., Lawrence, Kansas. 36 pp.

In this booklet, general instructions from a qualified printer of scientific publications are given to improve the quality of publication. Allen's emphasis is on photographs and he adequately covers special problems such as foldout illustrations, color plates, 300-line illustrations, and mathematical equations and chemical formulas. Good examples of results of various reductions are included.

Anonymous. 1978. *How to Plan Printing.* S. D. Warren Company, 225 Franklin Street, Boston, Massachusetts 02101.

This authorless booklet, written by a paper company, provides a good overview of the business of publishing, with chapters on writing, copy preparation, design, typesetting, selection of paper, printing, and binding. It is well written and is very effectively illustrated.

Blaker, Afred A. 1977. *Handbook for Scientific Photography.* W. H. Freeman & Co., San Francisco. 319 pp.

This is a well illustrated and thorough manual on all aspects of scientific photography, beginning with the basics of film types and printing, and discussing in depth the techniques for three-dimensional objects, photomicrography, and more. A brief section is included on photography of plant specimens. Relevant Kodak publications are listed in the bibliography. Although the book contains more than most botanists need to know, it is well organized so that it can serve as a reference text for most scientific photography problems. The companion text, *Field Photography: Beginning and Advanced Techniques* (1976) may also be useful to the botanist.

CBE Scientific Illustration Committee. In preparation. *Illustrating Science.* Council of Biology Editors, Inc., 9650 Rockville Pike, Bethesda, Maryland 20814.

Illustrating Science is expected to be published in the fall of 1986. The book will likely serve as the standard reference for all aspects of scientific illustration, with comprehensive detail. It will include discussions on computer graphics, reproduction in color, and an extensive section on reproduction of photographs.

CBE Style Manual Committee. 1983. *CBE Style Manual: A Guide for Authors, Editors, and Publishers in the Biological Sciences.* 5th ed. Council of Biology Editors, Inc., 9650 Rockville Pike, Bethesda, Maryland 20814. 324 pp.

This standard reference manual is useful for all phases of manuscript preparation and has a chapter on illustrative material.

Craig, James. 1974. *Production for the Graphic Designer.* Watson-Guptill Publications, New York. 206 pp.

Craig discusses the history and processes of typesetting, printing, folding, binding, and papermaking. Although the book is simplified, it gives a good view of the range of processes involved in the production of printed material. It is well illustrated and includes an extensive glossary.

Craig, James. 1978. *Phototypesetting: A Design Manual.* Watson-Guptill Publications, New York. 224 pp.

A good source of information, especially for authors preparing camera-ready text with the word processor.

Day, Robert A. 1983. *How to Write and Publish a Scientific Paper.* 2nd ed. ISI Press, 3501 Market Street, Philadelphia, Pennsylvania 19104. 181 pp.

Day has a chapter on "How to prepare effective illustrations," and we have already referred to his "How to design effective tables." It is an excellent book that even the most accomplished and accommodating publishing botanists could read with profit. Artists may find the book interesting in terms of how illustrations fit into the format of a scientific paper.

Downey, John C. & James L. Kelly. 1982. *Biological Illustration: Techniques and Exercises.* Iowa State University Press, Ames, Iowa. 128 pp.

The authors provide a series of exercises in biological illustrating for developing basic skills with pen and ink. There are chapters on the use of pen and pencil, perspective, shading, and use of overlays. This is a good manual for the beginning student of illustration who is willing to practice the exercises. An annotated list of supplies is included.

Gates, David. 1969. *Lettering for Reproduction.* Watson-Guptill Publications, New York. 192 pp.

Gates gives an interesting history of development of the Latin alphabet and explains the proportions and geometry of design of styles and spacing. He discusses useful tools and materials for hand lettering. The concepts apply to design and layout of lettering done from transfer type.

Hodges, Elaine R. S., editor. In preparation. *Guild Handbook of Scientific Illustration.* Van Nostrand Reinhold Co., New York.

The Guild of Natural Science Illustrators is producing a comprehensive text on scientific illustration. Drawing techniques and handling of specimens are treated by illustrators in the various fields including botany, invertebrate zoology, and archeology. Other planned chapters include business practices, tools, optics, printing, and conservation. The book will be illustrated with nearly 700 illustrations, including eight color pages. With an expected publication date of late 1986, this book will be a valuable, although expensive, text.

MacGregor, A. Jean. 1981. *Graphics Simplified: How to Plan and Prepare Effective Charts, Graphs, Illustrations, and Other Visual Aids.* University of Toronto Press, Toronto. 64 pp.

MacGregor offers a guide to the planning and preparation of graphs with emphasis on presentation visuals. She discusses basic types of graphs and considers the kinds of data each is best suited to. Her rules for good graphics are presented with yes and no situations. Included are instructions for the preparation of artwork for television, 35-mm slides, overhead transparencies, film strips, and poster displays.

Snyder, John. 1973. *Commercial Artist's Handbook.* Watson-Guptill Publications, New York. 264 pp.

The information is organized as an expanded glossary with the alphabetical entries cross-referenced with virtually no redundancy. The entries include furniture, tools, materials, media, generic terms for tools, and names of processes and methods. The items are described and often illustrated; their use is explained with helpful comments on the "Common difficulties" encountered.

Stone, B. & A. Eckstein. 1983. *Preparing Art for Printing.* Revised ed. Van Nostrand Reinhold Co., New York. 195 pp.

The book provides a good understanding of the processes of plate-making, printing, and typography. Also included is a good background for the preparation of art for publication and descriptions of equipment, tools, and materials available.

West, Keith. 1983. *How to Draw Plants: The Techniques of Botanical Illustration.* The Herbert Press, Ltd., London. 152 pp.

Botanical artists and students will learn useful techniques for handling plants and rendering accurate and artistic representations of them from living and preserved specimens. There are detailed chapters on pencil, ink, scraper (scratch) board, watercolor, acrylics, and photography, each organized into sections on equipment and techniques. The book is well illustrated with the author's own work. The chapter on plants in detail is packed with helpful hints toward acquiring proper perspective and the illusion of three-dimensions, with specific discussions on leaves, surface features, roots, fruits, seeds, and flowers.

Wood, Phyllis. 1979. *Scientific Illustration.* Van Nostrand Reinhold Co., New York. 148 pp.

Wood discusses preparation of layout, mechanicals, and printing, and offers a brief career guide. Halftone, black and white, and color illustrations are explained and well illustrated with student and professional examples. The book has no botanical orientation, but it provides a good introduction for the student artist, providing many examples to learn from in a variety of styles and techniques.

Zweifel, F. W. 1961. *A Handbook of Biological Illustration.* University of Chicago Press, Chicago. 130 pp.

The book is "for the biologist who is not an artist and the artist who is not a biologist." It is a good introduction to scientific publication and serves as a manual for students learning the trade on their own. Techniques for drawing in perspective and achieving tone are clearly discussed. The graphs and maps section is well illustrated. The text is basic without ever being elementary, and surprisingly complete for such a small book. The techniques are classic enough not to be outdated.

Literature Cited

Allen, A. 1977. Steps toward better scientific illustrations. 2nd ed. Allen Press, Inc., Lawrence, Kansas.

Alverson, W. S. 1984. *Quararibea pumila* (Bombacaceae), a new endemic from Costa Rica. Brittonia **36:** 252–256.

Anonymous. 1978. How to plan printing. S. D. Warren Co., Boston, Massachusetts.

Barkley, T. M. and others. 1977. Atlas of the flora of the Great Plains. Iowa State University Press, Ames, Iowa.

Barneby, R. C. 1964. Atlas of North American *Astragalus.* Mem. New York Bot. Gard. **13:** 1–1188.

Blaker, A. A. 1976. Field photography: Beginning and advanced techniques. W. H. Freeman & Co., San Francisco.

———. 1977. Handbook for scientific photography. W. H. Freeman & Co., San Francisco.

Brouwer, O. 1983. The cartographer's role and requirements. Scholarly Publishing **14:** 231–242.

CBE Style Manual Committee. 1983. CBE Style Manual: A guide for authors, editors, and publishers in the biological sciences. 5th ed. Council of Biology Editors, Inc., Bethesda, Maryland.

Craig, J. 1974. Production for the graphic designer. Watson-Guptill Publications, New York.

———. 1978. Phototypesetting: A design manual. Watson-Guptill Publications, New York.

Crawford, T. 1978. The visual artist's guide to the new copyright law. Graphic Artists Guild, New York.

Davis, P. H. & V. H. Heywood. 1963. Principles of angiosperm taxonomy. Oliver & Boyd, Edinburgh and London.

Day, R. A. 1983. How to write and publish a scientific paper. 2nd ed. ISI Press, Philadelphia, Pennsylvania.

Downey, J. C. & J. L. Kelly. 1982. Biological illustration: Techniques and exercises. Iowa State University Press, Ames, Iowa.

Gates, D. 1969. Lettering for reproduction. Watson-Guptill Publications, New York.

Gentry, A. 1980. Flora neotropica news. Taxon **29:** 379–381.

Griffin, J. R. & W. B. Critchfield. 1972. The distribution of forest trees in California. U.S.D.A. Forest Serv. Res. Pap. **PSW-82:** 1–114.

Holmgren, N. H. 1979. New penstemons (Scrophulariaceae) from the intermountain region. Brittonia **31:** 217–242.

Irwin, H. S. & R. C. Barneby. 1985. A new species of *Senna* (Caesalpiniaceae) from coastal northern Bahia, Brazil. Brittonia **37:** 192–194.

Johnston, D. F. 1982. Copyright handbook. 2nd ed. R. R. Bowker Co., New York.

Kallunki, J. A. 1976. Population studies in *Goodyera* (Orchidaceae) with emphasis on the hybrid origin of *G. tesselata*. Brittonia **28:** 53–75.

Keller, S. 1979. A revision of the genus *Wislizenia* (Capparidaceae) based on population studies. Brittonia **31:** 333–351.

Luteyn, J. L. 1983. Ericaceae—Part I. *Cavendishia*. Flora Neotropica **35:** 1–290.

MacGregor, A. J. 1981. Graphics simplified: How to plan and prepare effective charts, graphs, illustrations, and other visual aids. University of Toronto Press, Toronto.

May, P. H., A. B. Anderson, M. J. Balick & J. M. F. Frazão. 1985. Subsistence benefits from the babassu palm (*Orbignya martiana*). Economic Bot. **39:** 113–129.

Mori, S. A. & J. J. Pipoly. 1984. Observations on the big bang flowering of *Miconia minutiflora* (Melastomataceae). Brittonia **36:** 337–341.

Prance, G. T. & G. B. Schaller. 1982. Preliminary study of some vegetation types of the Pantanal, Mato Grosso, Brazil. Brittonia **34:** 228–251.

Robinson, A. H., R. D. Sale & J. L. Morrison. 1978. Elements of cartography. 4th ed. John Wiley & Sons, Inc., New York.

Snyder, J. 1973. Commercial artist's handbook. Watson-Guptill Publications, New York.

Stearn, W. T. 1951. Mapping the distribution of species. Pages 48–64 *in* J. E. Lousley, ed., The study of the distribution of British plants. Botanical Society of the British Isles, Oxford.

Stone, B. &. A. Eckstein. 1983. Preparing art for printing. Revised ed. Van Nostrand Reinhold Co., New York.

Voss, E. G., chairman editorial committee. 1983. International code of botanical nomenclature. Regnum Veg. **111:** 1–472.

West, K. 1983. How to draw plants: The techniques of botanical illustration. The Herbert Press, Ltd., London.

Wood, P. 1979. Scientific illustration. Van Nostrand Reinhold Co., New York.

Zweifel, F. W. 1961. A handbook of biological illustration. University of Chicago Press, Chicago.

Index

NOEL H. HOLMGREN and BOBBI ANGELL

Noel Herman Holmgren was born in Salt Lake City, Utah in 1937. He grew up in Logan, Utah with the intention of becoming an artist, but after two years of college as an art major his interest in field botany overtook his dedication to art. In 1962 he obtained a B.S. degree in botany at Utah State University where he took most of the courses taught by his father in plant taxonomy. In the summer of 1961, after spending five weeks collecting plants with Arthur Cronquist in southeastern Utah, Noel decided to pursue his Ph.D. studies under his direction as a Graduate Assistant at the New York Botanical Garden. Noel received his Ph.D. from Columbia University in 1968 and joined the Garden's staff to work with Cronquist on the Intermountain Flora project and on his specialty, the taxonomy of the family Scrophulariaceae. In 1969 he married Patricia Kern. Noel's interest in art has never left him. As editor of *Brittonia* since 1977 he has been especially critical of artwork submitted for publication.

Bobbi Angell was born in Oil City, Pennsylvania in 1955. Her time growing up was divided between Mamaroneck, New York and southern Vermont. A strong leaning towards the latter led to a special interest in nature. Her undergraduate career began at Prescott College in Arizona where she began studying natural history. Her professors steered her towards illustrating as a means of learning desert plants most accurately. She transferred to the University of Vermont to major in botany, but soon allowed her skills at drawing to overshadow her abilities in science. While completing her B.S. degree, she developed the idea of making a living as a botanical illustrator. After exploring herbaria across the country, she began her career as an artist on the Intermountain Flora project at the New York Botanical Garden in 1978, where she has remained and now resides as the staff freelance artist. Other freelance jobs have included field guides, private watercolor commissions, and product and giftware design.